SpringerBriefs in Electrical and Computer Engineering

More information about this series at http://www.springer.com/series/10059

Cam Nguyen · Meng Miao

Design of CMOS RFIC Ultra-Wideband Impulse Transmitters and Receivers

 Springer

Cam Nguyen
Texas A&M University
College Station, TX
USA

Meng Miao
Texas A&M University
College Station, TX
USA

ISSN 2191-8112   ISSN 2191-8120   (electronic)
SpringerBriefs in Electrical and Computer Engineering
ISBN 978-3-319-53105-2   ISBN 978-3-319-53107-6   (eBook)
DOI 10.1007/978-3-319-53107-6

Library of Congress Control Number: 2017931998

Printed on acid-free paper

This Springer imprint is published by Springer Nature
The registered company is Springer International Publishing AG
The registered company address is: Gewerbestrasse 11, 6330 Cham, Switzerland

Preface

Impulse-based ultra-wideband (UWB) systems, notably unlicensed UWB systems operating across or within 3.1–10.6 GHz, are unique systems possessing many desired characteristics, owing to the transmission and reception of only a single signal having pulse waveform at all times, rather than multiple consecutive CW signals having sinusoidal waveform at different times as in CW-based systems. This unique operation is equivalent to transmitting and receiving many CW signals across an extremely broad bandwidth concurrently. A UWB impulse system can hence be considered to some extent as electrically equivalent to multiple CW systems, each operating at a single frequency in an ultra-wide bandwidth, working simultaneously. It is this unique operation that makes the UWB impulse systems distinctly different from the UWB CW systems, not only in the design and performance, but also in applications. The three most important subsystems in a UWB impulse system, as in any other RF systems, are the UWB transmitter, receiver, and antenna.

Silicon-based Complementary-Metal-Oxide-Semiconductor (CMOS) RFICs plays a dominant role in realizing miniature low-cost and low-power consumption single-chip RF systems, which are particularly essential for portable or hand-held implementations. CMOS single-chip RFIC transmitters and receivers are especially needed for advanced miniature UWB impulse systems.

This book is devoted to the design of CMOS RFIC UWB impulse transmitters and receivers and their components for UWB impulse systems. Specifically, it addresses three main topics of UWB impulse transmitters and receivers: CMOS UWB transmitter design, CMOS UWB receiver design, and UWB uni-planar antenna design. The book also describes the actual design, simulation, fabrication and measurement of these subsystems, which can readily be integrated together to realize UWB impulse systems for potential applications in communications and sensing.

Although the book is succinct, the material is very much self-contained and contains practical, valuable and sufficient information presented in such a way that allows readers with undergraduate background in electrical engineering or physics, with some experiences or graduate courses in RF circuits, to understand and design easily UWB impulse transmitters, receivers, and antennas for various UWB applications.

The book is useful for engineers, physicists, and graduate students who work in radar, sensor, and communication systems as well as those involved in the design of RF circuits and systems. It is our sincere hope that the book can serve not only as a reference for the development of UWB systems and components, but also for possible generation of innovative ideas that can benefit many existing sensing and communication applications or be implemented for other new applications.

College Station, USA   Cam Nguyen
Chandler, USA   Meng Miao

Contents

Chapter 1
Introduction

The origin of "Ultra-Wideband" technology can be traced back to the early 1960s on time-domain electromagnetics, when the study of electromagnetic-wave propagation was primarily viewed from the time-domain perspective, rather than from the more common sinusoidal-signal based frequency domain. Nevertheless, the term Ultra-Wideband (UWB) was only used for the first time in 1994 by the U.S. Department of Defense to indicate impulse-based systems, which by the nature of the pulse waveforms, involve very wide bandwidths. In general, however, ultra-wideband (UWB) systems should not be limited to systems employing pulse signals and/or over operating over a specific bandwidth or frequency range; rather they should refer to systems that operate over an ultra-wide bandwidth regardless of the employed signal waveforms (e.g., pulse or sinusoid). These systems may have different architectures and be used for different applications such as impulse-based systems and continuous-wave (CW) sinusoid-based systems with different transmitting signal's frequency modulations like frequency modulated CW (FMCW) or stepped-frequency system. Specifically, as commonly acknowledged, a system may be classified as a UWB system when its operating frequency range is wider than 500 MHz or 20% fractional bandwidth.

The most distinguished difference between a UWB impulse[1] system and a UWB FMCW or stepped-frequency system (and in general other non-pulse or CW based systems) is their transmitting waveforms. A UWB FMCW system transmits and receives CW sinusoidal signals, one signal at each frequency, subsequently across a wide bandwidth. A UWB FMCW system does not transmit and receive signals of different frequencies simultaneously. That is, a UWB FMCW system operates over a wide bandwidth of single-frequency signals. A UWB stepped-frequency system transmits a train of CW signals across a wide frequency range, each at different frequency separated by a certain amount, toward targets and receives reflected signals from the targets. The received in-phase and quadrature signals are then

---

[1]The word "impulse" is used loosely here. It does not indicate a "true" impulse, but simply pulses with a narrow pulse width.

© The Author(s) 2017
C. Nguyen and M. Miao, *Design of CMOS RFIC Ultra-Wideband Impulse Transmitters and Receivers*, SpringerBriefs in Electrical and Computer Engineering, DOI 10.1007/978-3-319-53107-6_1

transformed into a synthetic pulse in time domain using inverse discrete Fourier Transform. A UWB stepped-frequency system has a very narrow instantaneous bandwidth at each frequency, hence resulting in high signal-to-noise ratio at the receiver. Its entire bandwidth, on the other hand, can be very wide, thereby leading to fine resolution. Moreover, its high average transmitting power enables deep penetration or long range. Although the final received signal is transformed into a time-domain pulse signal, it is still CW-based and does not transmit and receive signals of different frequencies simultaneously. On the other hand, a UWB impulse system transmits and receives a periodic (non-sinusoidal) impulse-type signal, which contains many constituent signals occurring simultaneously, each having a different frequency. In other words, a UWB impulse system "concurrently" transmits and receives many sinusoidal signals having different frequencies. A typical UWB impulse system transmits and receives information using millions of narrow pulses each second with extremely low-power spectral densities across an ultra-wide band spectrum. It is this characteristic that makes UWB impulse system and UWB FMCW or stepped-frequency system (and other CW based systems) distinctively different not only in their design, operation and performance, but also in their possible applications. UWB impulse systems have many advantages as mentioned in Chap. 2, including: fine range resolution, long range (to some extent), minimum interference with and from other existing signals or co-operating RF systems, high multi-path resolution, low probability of interception or detection, reduced signal fading, improved accuracy of locating and tracking, and simple and low-cost architecture.

UWB systems find numerous applications for military, security, civilian, commerce and medicine, and various UWB components and systems have been developed for different applications, e.g., [1–26]. They have received significant interests, especially after the FCC's Notice of Inquiry in 1998 [27] and Report and Order in 2002 [28] for unlicensed uses of UWB devices within the 3.1–10.6 GHz frequency band. Particularly, UWB techniques implemented from 3.1 to 10.6 GHz are promising for high data-rates according to the Shannon's theorem [29], which states that the channel capacity characterized by the highest data rate of reliable transmission in bits per second (bps) is directly proportional to the channel bandwidth, and short-range ad hoc networking and communications, such as home networks, in-building communications and data transmission, and cordless phones.

UWB systems operating from 3.1 to 10.6 GHz are typically implemented with two widely used architectures. One is the orthogonal frequency division multiplexing (OFDM) based multi-band scheme, in which the frequency band from 3.1 to 10.6 GHz is divided into tens of hundred-MHz bands and CW sinusoids are used as the transmitting signals. The other is the impulse-based scheme, which employs impulse-like transmitting signals. In the multi-band OFDM UWB systems, data are transmitted using OFDM on different bands in a time-interleaved fashion. Each band has a minimum bandwidth of 500 MHz and frequency hopping is employed to cover a wide bandwidth within or across 3.1–10.6 GHz. A multi-band device can be designed to dynamically select which bands to be used for transmission; therefore, the multi-band OFDM UWB scheme has the advantages of inherent robustness to multi-path fading, excellent robustness to narrowband interference, and ease in complying with the worldwide regulations for UWB communications. As compared to multi-band OFDM UWB systems, UWB impulse systems have a simpler architecture and circuit due to the facts that they only require a simple pulse generator working as the transmitter and a simple sampling circuit as the receiver. Furthermore, a UWB impulse system can use completely the ultra-wide 7.5-GHz bandwidth from 3.1 to 10 GHz, hence possibly reducing the chance of fading in situations where noise exists in a narrow frequency band within the UWB band, which results in better immunity to destructive channel environments. This is an advantage as compared to the OFDM case, where the noise in a particular frequency channel may disrupt that channel and ultimately disturbs the system's performance across the operating bandwidth. Additionally, since an impulse UWB system uses an impulse signal having a very short duration, the accuracy of position detection and tracking is higher.

Various UWB impulse systems and components were developed using discrete microwave components and discrete microwave hybrid circuits, generally known as microwave integrated circuits (MIC), e.g., [1–17]. These systems and components are not very compact and relatively expensive. Miniaturized and low-cost UWB impulse systems and components are needed to reduce the size and lower the cost so that many UWB impulse systems can be placed in a small space with affordable cost for certain applications. One of the most attractive technologies for realizing such compact UWB impulse systems and components is the silicon-based radio frequency integrated circuit (RFIC).

Silicon-based Complementary-Metal-Oxide-Semiconductor (CMOS) RFICs and systems have advanced significantly and can perform at very high frequencies. CMOS RFICs have lower cost and better abilities for direct integration with digital ICs (and hence better potential for complete system-on-a-chip) as compared to those using III–V compound semiconductor devices. CMOS RFICs are also small and have low power dissipation, making them suitable for battery-operated systems. CMOS RFICs are thus very attractive for RF systems and, in fact, the principal choice for commercial wireless markets. CMOS single-chip RFIC transmitters and receivers are therefore needed for advanced miniature UWB systems. As a comparison, a typical MIC transmitter for UWB systems may have a size of around 50 mm × 100 mm, while a CMOS RFIC transmitter (as designed for UWB impulse systems and included in this book) could be realized with a size smaller than 1 mm × 1 mm. It is also further noted that single-chip transmitters and receivers, and ultimately single-chip communication and sensing systems, are highly desirable for many commercial and military applications. Recently, various CMOS RFICs have been developed for UWB systems [18–25].

Since UWB impulse signals cover a relatively wide frequency band with a fractional bandwidth typically exceeding 25%, there is great challenge in the design of antennas for UWB impulse systems. As these UWB antennas essentially operate in the time domain and radiate all frequency components simultaneously, they not only need to radiate the energy over a wide frequency range, but also require a linear phase response over the entire frequency band to avoid signal distortion [30].

The linear-phase requirement demands an antenna with a fixed phase center at different frequencies in the band. However, many of existing wideband antennas, such as the tapered slot and log periodical antennas, have floating phase centers at different frequencies and hence are not suitable for UWB impulse systems. Low-cost, compact, easy-to-manufacture coplanar UWB antennas that are omni-directional, radiation-efficient and have a stable UWB response are desirable for UWB impulse systems. These coplanar antennas can be easily integrated with UWB CMOS RFIC transmitter and receiver chips.

This book presents the design of CMOS RFIC UWB impulse transmitters and receivers and their components for UWB impulse systems. It is particularly noted that there are many aspects in the design of these subsystems and a complete coverage would require a book of substantial volume. The objective of this book is not to provide a full coverage of CMOS RFIC UWB impulse transmitters and receivers or the design and theory of UWB impulse systems. It is our intention to address only the essential parts of these CMOS RFIC transmitters, receivers and their components including design, analysis, and measurement in a concise manner. The book provides sufficient and specific design information of CMOS RFIC transmitters and receivers for UWB impulse systems to enable readers to understand and design them for their intended applications, whether for research or for commercial usage. Specifically, it addresses three main topics of UWB impulse transmitters and receivers: transmitter design, receiver design, and antenna design. The book also describes the actual design, simulation, fabrication and measurement of these subsystems, which can readily be integrated together to realize UWB impulse systems for potential applications in short-range communications, surface and short-range subsurface sensing, and short-range detection and tracking. It is hoped that this book would contribute to the design of CMOS RFIC UWB transmitters and receivers for UWB impulse systems, a topic that is not addressed in regular books written on the design of transmitters, receivers and systems based on conventional microwave circuits.

It is particularly noted that the UWB impulse transmitters, receivers and components addressed in this book are within the context of the unlicensed UWB systems operating across or within 3.1–10.6 GHz. Nevertheless, all the design, analysis and measurement of these subsystems and their components presented in this book are applicable to other operating frequency ranges outside the 3.1–10.6 GHz for other UWB impulse systems, as well as some UWB CW systems, and applications.

One particular note needed to be mentioned here is that the design of RFICs, including those constituting the UWB impulse transmitters and receivers presented in this book, is a rather complicated process involving various steps, which is beyond the scope of this book. The essence of RFIC design including the design fundamentals, lumped and distributed elements for RFIC, analyses, simulations, layouts, post-layout simulations and optimizations, and measurements can be found in [30].

The book contains six chapters. Chapter 1 gives an introduction and background of UWB systems for wireless communications and sensing. Chapter 2 discusses the basics of UWB impulse systems including system architecture, advantages and applications, signals, basic modulations, transmitter and receiver frontends, and antennas. Chapter 3 addresses the design of UWB transmitters including an overview of basic components, design of impulse and monocycle pulse generators, Bi-Phase Shift-Keying (BPSK) modulator design, and design of a UWB CMOS RFIC tunable transmitter front-end. Chapter 4 presents the design of UWB receivers including the design of UWB low-noise amplifiers, strobe pulse generator, correlators, and a UWB CMOS RFIC receiver front-end. Chapter 5 presents the design, fabrication and measurement of a compact UWB uniplanar antenna suitable for UWB impulse systems. It also covers the measurement results of a UWB transmit prototype that integrates the UWB monocycle-pulse generator and the UWB uniplanar antenna. Finally, Chap. 6 gives the summary and conclusion.

References

1. M.I. Skolnik, *An Introduction to Impulse Radar*, Naval Research Laboratory, Washington, DC, NRL Memorandum Report 6755, Nov 1990
2. D.J. Daniels, *Surface Penetrating Radar* (IEE Press, London, U.K., 1996)
3. D.J. Daniels, D.J. Gunton, H.F. Scott, Introduction to subsurface radar. IEE Proceedings **135** (4), 278–320 (1988)
4. J.D. Taylor, *Introduction to Ultra-Wideband Radar Systems* (CRC Press, Boca Raton, FL, 1995)
5. J.D. Taylor, *Ultra-Wideband Radar Technology* (CRC Press, Boca Raton, FL, 2001)
6. R.J. Fontana, Recent applications of ultra wideband radar and communications systems, in *Ultra-Wideband, Short-Pulse Electromagnetics 5*, ed. by P.D. Smith, S.R. Cloude (Kluwer Academic/Plenum Publishers, New York, 2002), pp. 225–234
7. A. Yarovoy, L. Ligthart, Full-polarimetric video impulse radar for landmine detection: experimental verification of main design ideas, in *Proceedings of 2nd International Workshop on Advanced Ground Penetrating Radar* (2003), pp. 148–155
8. J.S. Park, C. Nguyen, An ultra-wideband microwave radar sensor for nondestructive evaluation of pavement subsurface. IEEE Sens. J. **5**, 942–949 (2005)
9. J.S. Lee, C. Nguyen, T. Scullion, A novel compact, low-cost impulse ground penetrating radar for nondestructive evaluation of pavements. IEEE Trans. Instrum. Measur. **IM-53**, 1502–1509 (2004)
10. S. Azevedo, T.E. McEwan, Micropower impulse radar. Sci. Tech. Rev. 17–29 (1996)
11. A. Batra et al. Design of a multiband OFDM system for realistic UWB channel environments. IEEE Trans. Microwave Theory Tech. **MTT-52**(9), 2123–2138 (2004)
12. J.W. Han, C. Nguyen, Development of a tunable multi-band UWB radar sensor and its applications to subsurface sensing. IEEE Sens. J. **7**(1), 51–58 (2007)
13. R.J. Fontana, Recent system applications of short-pulse ultra-wideband (UWB) technology. IEEE Trans. Microwave Theory Tech. **MTT-52**, 2087–2104 (2004)
14. J.W. Han, C. Nguyen, Coupled-Slotline-Hybrid sampling mixer integrated with step-recovery-diode pulse generator for UWB applications. IEEE Trans. Microwave Theory Tech. **MTT-53**(6), 1875–1882 (2005)
15. J.W. Han, C. Nguyen, On the development of a compact sub-nanosecond tunable monocycle pulse transmitter for UWB applications. IEEE Trans. Microwave Theory Tech. **MTT-54**(1), 285–293 (2006)

16. J.S. Lee, C. Nguyen, Novel low-cost ultra-wideband, ultra-short-pulse transmitter with MESFET impulse-shaping circuitry for reduced distortion and improved pulse repetition rate. IEEE Microwave Wirel. Compon. Lett. **11**(5), 208–210 (2001)
17. J. Han, C. Nguyen, Ultra-wideband electronically tunable pulse generators. IEEE Microwave Wirel. Compon. Lett. **14**(3), 112–114 (2004)
18. A. Bevilacqua, A.M. Niknejad, An ultrawide band CMOS low-noise amplifier for 3.1–10.6-GHz wireless receivers. IEEE J. Solid State Circuits **39**(12), 2259–2268 (2004)
19. R. Xu, Y. Jin, C. Nguyen, Power-efficient switching-based CMOS UWB transmitters for UWB communications and radar systems. IEEE Trans. Microwave Theory Tech. **MTT-54**(8), 3271–3277 (2006)
20. X. Guan, C. Nguyen, Low-power-consumption and high-gain CMOS distributed amplifiers using cascade of inductively coupled common-source gain cells for UWB systems. IEEE Trans. Microwave Theory Tech. **MTT-54**(8), 3278–3283 (2006)
21. M. Miao, C. Nguyen, On the development of an integrated CMOS-based UWB Tunable–Pulse transmit module. IEEE Trans. Microwave Theory Tech. **MTT-54**(10), 3681–3687 (2006)
22. M. Miao, C. Nguyen, Fully integrated CMOS impulse UWB transmitter front-ends with BPSK modulation. Microwave Opt. Technol. Lett. **52**(7), 1609–1614 (2010)
23. R. Xu, C. Nguyen, An ultra-wideband low power-consumption low noise-figure, high-gain, RF-power efficient DC-3.5-GHz CMOS integrated sampling mixer. IEEE Trans. Microwave Theory Tech. **MTT-56**, 1069–1075 (2008)
24. X. Guan, C. Nguyen, A novel CMOS distributed receiver front-end for wireless ultra-wideband receivers. Microwave Opt. Technol. Lett. **52**(8), 1790–1792 (2010)
25. R. Xu, C. Nguyen, A high precision close-loop programmable CMOS delay generator for uwb and time domain RF applications. Microwave Opt. Technol. Lett. **53**(2), 390–392 (2011)
26. "Revision of part 15 of the commission's rules regarding ultra-wideband transmission systems." [Online]. FCC Notice of Inquiry, adopted 20 Aug 1998, released 1 Sept 1998. Available: http://www.fcc.gov/oet/dockets/et98-153
27. "Revision of part 15 of the commission's rules regarding ultra-wideband transmission systems," FCC Report and Order, adopted 14 Feb 2002, released 15 July 2002
28. C. Shannon, A mathematical theory of communication. The Bell Syst. Tech. J. **27**, 379–423, 623–656 (1948)
29. H.G. Schantz, Introduction to ultra-wideband antennas. IEEE Conf. Ultra Wideband Syst. Technol. 1–9 (2003)
30. C. Nguyen, *Radio-Frequency Integrated-Circuit Engineering* (Wiley, New York, 2015)

Chapter 2
Fundamentals of UWB Impulse Systems

2.1 Introduction

In this chapter, the basic concepts and features of UWB impulse systems are presented. Different types of UWB impulse signals and commonly used modulation topologies for UWB impulse systems are described. The chapter also provides the core topologies of UWB impulse transmitters and receivers as well as brief overview of the UWB antennas for UWB impulse systems. These are essential for the design and understanding of UWB impulse systems.

2.2 UWB Overview

2.2.1 UWB Basics

Over the unlicensed frequency range of 3.1–10.6 GHz, the FCC has defined UWB signals as those that occupy a 10-dB bandwidth of greater than 500-MHz bandwidth or larger than 20% fractional bandwidth as defined by

$$\text{Fractional Bandwidth} = \frac{2(f_H - f_L)}{f_H + f_L} \geq 20\% \tag{2.1}$$

where f_H and f_L are the upper and lower frequency limits, respectively, of the 10-dB bandwidth.

The FCC also requires that the power emission levels of the UWB signals within the UWB spectrum of 3.1–10.6 GHz must be sufficiently low to avoid interference with other existing communication systems, technologies and services operating in the same UWB allocated bands, hence enabling them to exist together. Specifically,

© The Author(s) 2017
C. Nguyen and M. Miao, *Design of CMOS RFIC Ultra-Wideband Impulse Transmitters and Receivers*, SpringerBriefs in Electrical and Computer Engineering, DOI 10.1007/978-3-319-53107-6_2

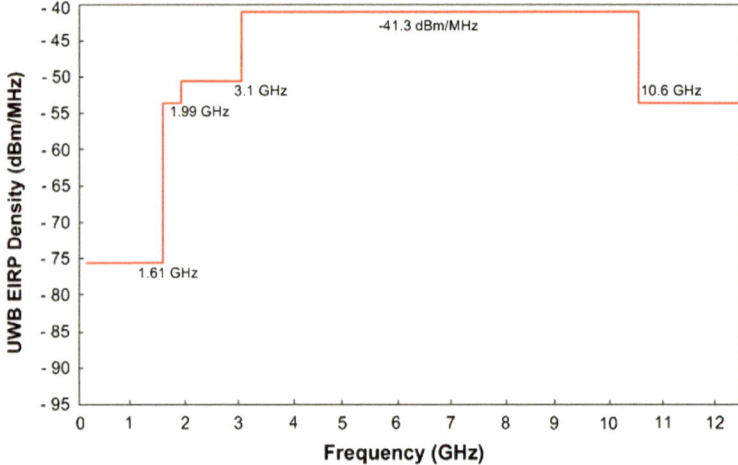

Fig. 2.1 Spectrum mask of the UWB effective isotropic radiated power (EIRP) from transmitting antenna

the FCC requires that the maximum allowed power spectral density (PSD) not to exceed −41.25 dBm/MHz as designated in the UWB spectrum mask shown in Fig. 2.1. This kind of power is low enough not to cause interference to other services, such as Wireless Fidelity (WiFi) operating under different rules that share the same bandwidth within the UWB frequency range. The limited emitting power presents a serious challenge to these unlicensed UWB systems because other RF systems or services sharing the same band of operation on licensed or unlicensed bands are likely to have a much higher transmitting power and, therefore, would subject the UWB receivers to considerable interference. This low RF emitting power requirement inevitably limits these unlicensed UWB systems to work only within short ranges, making it suitable to employ miniature CMOS RFICs whose RF power capability and dc power consumption are relatively small. It is noted that in UWB impulse systems, the transmitting pulse spreads the energy over a wide frequency band as compared to narrow-band signals as illustrated in Fig. 2.2.

2.2.2   UWB Advantages

Compared to CW based systems, UWB impulse systems have many advantages as following:

1. **Fine Resolution and Long Range**

   UWB impulse systems typically have much wider instantaneous bandwidths than those of CW based systems due to the extremely wideband nature of the

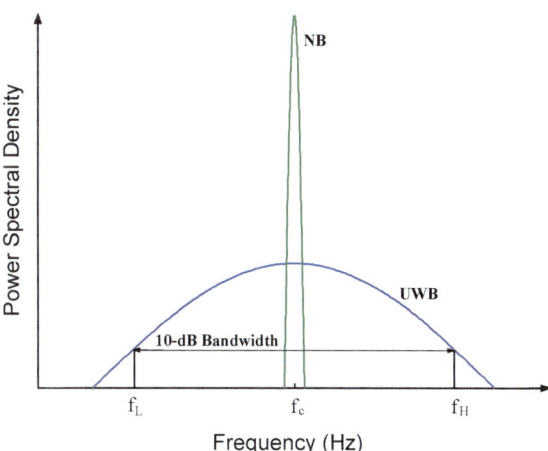

Fig. 2.2 Power level of a UWB signal with respect to a typical narrowband (NB) signal

impulse-type signals. These signals contain both low and high frequency components, making impulse systems very suitable for applications requiring fine range-resolution and/or long range. An ultra-wide bandwidth directly leads to fine range-resolution due to the fact that the range resolution is inversely proportional to the bandwidth. An ultra-wide frequency range spanning across low and high frequencies enables long range as compared to a CW frequency range containing only comparable high frequencies due to small attenuation at low frequencies and hence long propagating distance of low-frequency signals. It is noted that, while similar range-resolution and range can also be achieved for CW based systems operating with same bandwidth and frequencies, respectively, it is very difficult (if not impossible) to realize an extremely wide-band CW system.

2. **High Multi-Path Resolution and Low Interference with Other Existing Signals**

The transmitting energy spectral density of UWB impulse systems is typically much lower than that of CW based systems for the same input power since the total energy is spread over a wide range of frequencies of the impulse signal. This effectively produces much smaller interference to the signals of other existing or co-operating RF systems. Typical impulse signals have a very narrow pulse width and hence the transmission duration of impulse signals is very short in most cases. The signals returned from targets, in turn, have a very short time-window of opportunity to collide with the line-of-sight signals and less likely cause signal degradation, hence very high multi-path resolution can be achieved. The large frequency diversity of the ultra-wide spectrum of an impulse signal makes impulse signal relatively resistant to intentional and unintentional jamming or interference, because it is difficult to jam every frequency component in the ultra-wideband spectrum at once. Even if some of the frequency components are jammed, there is still possibly a large range of frequencies that remains untouched. UWB impulse systems offer excellent immunity to interference from other existing signals, while also causing minimum interference to these signals.

3. Low Probability of Interception or Detection

The lower-energy spectrum density of impulse signals also makes unintended detection more difficult than CW systems, hence resulting in low probability of interception or detection, which is desirable for secure and military applications.

4. Reduced Signal Diminishing

The ultra-wide bandwidth of UWB impulse systems leads to high frequency diversity that reduces the chance of signal diminishing in certain operating environments with severe multi-path fading at some frequencies, such as indoors, urban settings or mountainous terrains, or when signal attenuation at some frequencies are excessively high, which hinder the sensing capabilities, or where noise exists in a narrow-frequency range within the operating band, resulting in better immunity to destructive environments.

5. Reduced Signal Diminishing

The accuracy of locating and tracking for UWB impulse systems is higher due to the narrow time-duration of pulse signals typically used for transmission which causes much better timing precision than CW systems such as global positioning system (GPS).

6. Simple and Low-Cost Architecture

UWB impulse systems can be implemented with a simpler architecture than their wideband CW counterparts including multi-band OFDM systems, in which the core signal generator of the transmitter can be realized with a simple pulse generator without an up-conversion circuit and the core mixer of the receiver can be implemented with a simple direct-conversion (to baseband) sampling circuit, that does not require an intermediate Frequency-conversion stage, as compared to more complicated wideband signal source and mixer typically used in CW based systems. Moreover, a complex frequency synthesizer (or even simple oscillators) needed for the transmitter and receiver in CW systems is avoided. Additionally, a simple impulse UWB system with no frequency synthesizer or oscillator consumes less power, thereby extending the operating life, which is attractive for battery-operated portable devices.

UWB impulse systems of course also have several disadvantages as compared to CW systems. For instance, the receiver's noise figure of UWB impulse systems is much higher than that of CW systems, which in turn limits the sensitivity and hence the dynamic range of the receiver, preventing their use for applications requiring very high sensitivity and/or large dynamic range. It is also much more complicated to design antennas for UWB impulse systems due to wide bandwidths and the need to maintain signal fidelity across such bandwidths.

2.2.3 UWB Applications

UWB impulse systems find numerous applications for military, security, civilian, commerce and medicine. The following are some of the existing and emerging applications of UWB impulse systems:

Military and Security Applications: detection, location and identification of targets such as aircrafts, tunnels, concealed weapons, hidden illegal drugs, buried mine and unexploded ordnance (UXO); locating and tracking personnel; detection and identification of hidden activities; access control; through-wall imaging and surveillance; building surveillance and monitoring.

Civilian and Commercial Applications: detection, identification and assessment of abnormal conditions of civil structures such as pavements, bridges, buildings, buried underground pipes; detection, location and identification of objects; asset and inventory management; radio-frequency identification (RFID); monitoring of personal properties such as cars, homes and valuable items; intrusion detection; asset tracking; measurement of liquid volumes and levels; inspection, evaluation and process control of materials; geophysical prospecting, altimetry; collision and obstacle avoidance for automobile and aviation.

Medical Applications: detection and imaging of tumors; health monitoring of elders; health examination of patients; medical imaging.

It is particularly noted that the major application of UWB impulse systems operating within 3.1–10 GHz is for short-range communication due to its inherently very high data transfer rate for short distances. UWB impulse systems can send and receive high-speed data with very low power at relatively low cost and hence are attractive for short-range wireless communication areas. Specifically, the UWB technologies primarily target indoor short-range high-bit-rate applications such as home networking, high-speed wireless local area network (LAN), and personal area network (PAN) communications.

2.3 UWB Impulse Signals

The selection of impulse-signal types for UWB impulse systems is one of the fundamental considerations in designing UWB impulse systems, antennas, and circuits because the type of an impulse determines the UWB signal's spectrum characteristic. Many types of impulse signals such as step pulse, Gaussian-like (or monopolar) impulse, Gaussian-like single-cycle (or monocycle) pulse, Gaussian-like doublet pulse, and multi-cycle pulse can be used for UWB impulse systems. Among those, Gaussian-like impulse, doublet pulse, and monocycle pulse are typically used in UWB impulse systems. Particularly, the monocycle pulse is preferred in most UWB impulse systems because of its spectral characteristics (having no dc) that facilitate easier wireless transmission than the impulse, wider bandwidth than the multi-cycle pulse, and easier to realize than the doublet pulse.

2.3.1 Gaussian Impulse

Figure 2.3 shows the time-domain waveform of a Gaussian impulse that has a shape of the Gaussian distribution, along with its frequency-domain waveform or spectral response. The impulse is assumed to have 200-ps pulse duration (or pulse width). The Gaussian impulse can be expressed as

$$y(t) = Ae^{-a^2 t^2} \tag{2.2}$$

where A is the maximum amplitude of the Gaussian impulse and a is the constant that determines the slope of the Gaussian pulse. The spectral response containing the spectral components of the Gaussian impulse is obtained by taking its Fourier transform as

$$Y(\omega) = \frac{A}{a\sqrt{2}} e^{-\frac{\omega^2}{4a^2}} \tag{2.3}$$

The frequency corresponding to the peak value of the impulse in the frequency domain is $f_o = 0$. The 3-dB bandwidth of the Gaussian impulse can be derived by letting the amplitude of the impulse at the 3-dB band-edge equal to the $1/\sqrt{2}$ of the maximum value at $f = 0$ as

$$\Delta f = 0.8326 \frac{a\sqrt{2}}{2\pi} \tag{2.4}$$

2.3.2 Gaussian Monocycle Pulse

Gaussian monocycle pulse is the first derivative of the Gaussian impulse signal. Figure 2.4 shows a Gaussian monocycle pulse having the same 200-ps pulse duration as the Gaussian impulse shown in Fig. 2.3 and its spectrum. The Gaussian monocycle pulse is described by

$$y(t) = -2a^2 At e^{-a^2 t^2} \tag{2.5}$$

The spectral response of the Gaussian monocycle pulse is given as

$$Y(\omega) = \frac{i\omega A}{a\sqrt{2}} e^{-\frac{\omega^2}{4a^2}} \tag{2.6}$$

The frequency corresponding with the peak value of the Gaussian monocycle pulse in the spectrum is obtained as

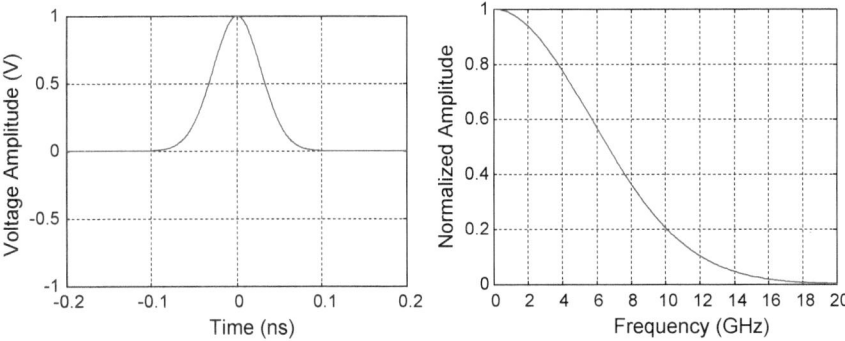

Fig. 2.3 Gaussian impulse with 200-ps pulse duration and its frequency spectrum

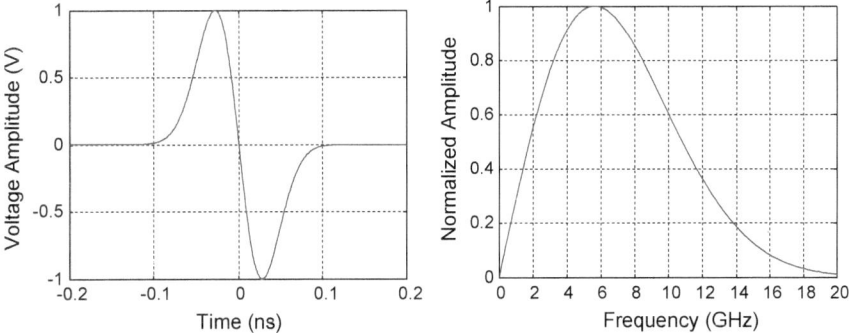

Fig. 2.4 Gaussian monocycle pulse with 200-ps pulse duration and its frequency spectrum

$$f_o = \frac{a\sqrt{2}}{2\pi} \tag{2.7}$$

and the 3-dB bandwidth can be derived as

$$\Delta f = 1.155 \frac{a\sqrt{2}}{2\pi} = 1.155 f_o = \frac{1.155}{T_p} \tag{2.8}$$

where $T_p = 1/f_o$ is the pulse duration, which shows that the 3-dB bandwidth of the Gaussian monocycle pulse is approximately equal to 115% of the pulse's center frequency f_o. Figures 2.5 and 2.6 show the waveforms and spectrums of various Gaussian monocycle pulses having different pulse durations.

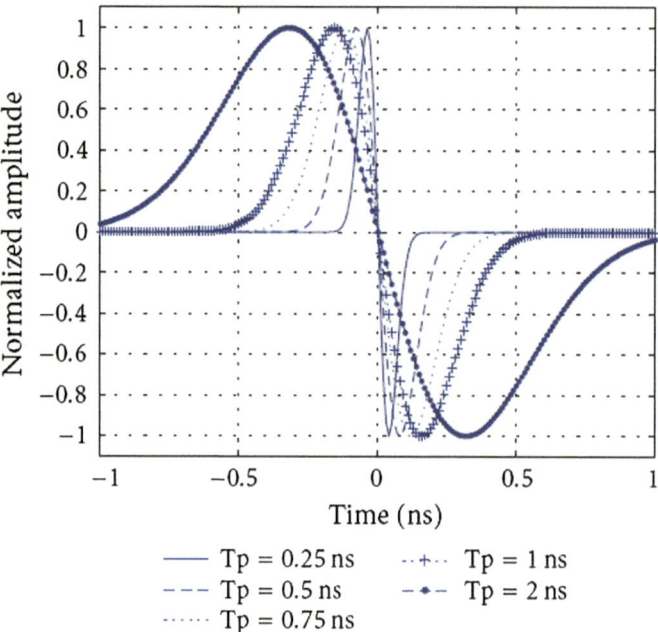

Fig. 2.5 Gaussian monocycle pulses with different pulse durations

Fig. 2.6 Spectrum of
Gaussian monocycle pulses
with different pulse durations

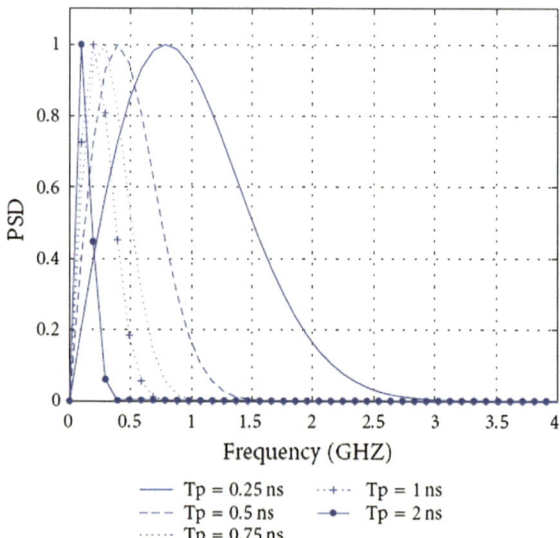

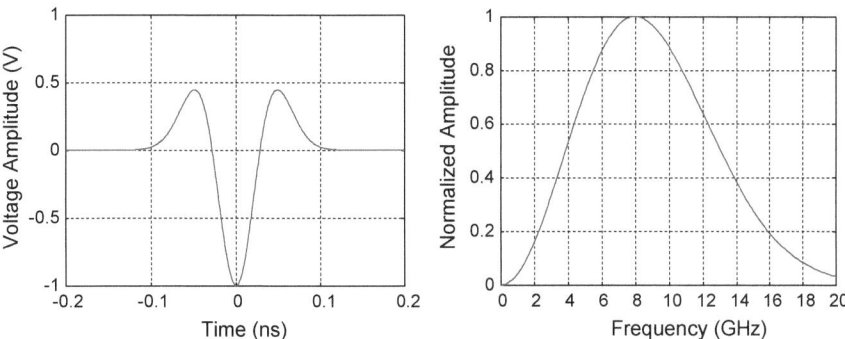

Fig. 2.7 Gaussian doublet pulse with 200-ps pulse duration and its frequency spectrum

2.3.3   Gaussian Doublet Pulse

Figure 2.7 shows a Gaussian doublet pulse having 200-ps pulse duration and its spectrum. The Gaussian doublet pulse is the second derivative of the Gaussian impulse signal and hence can be expressed as

$$y(t) = -2a^2 A e^{-a^2 t^2} (1 - 2a^2 t^2) \tag{2.9}$$

The spectral response of the Gaussian doublet pulse is

$$Y(\omega) = \frac{-A\omega^2}{a\sqrt{2}} e^{-\frac{\omega^2}{4a^2}} \tag{2.10}$$

The frequency at which the peak value of the Gaussian doublet pulse occurs in the spectrum is

$$f_o = \frac{a}{\pi} \tag{2.11}$$

This frequency is higher than that given in (2.7) for the Gaussian monocycle pulse. The 3-dB bandwidth can be derived as

$$\Delta f = 1.155 \frac{a\sqrt{2}}{2\pi} = 1.155 \frac{f_o}{\sqrt{2}} \tag{2.12}$$

Compared to the bandwidth of the Gaussian monocycle pulse given in (2.8), the absolute bandwidth of the Gaussian doublet pulse is same, yet the fractional bandwidth is larger assuming the same pulse duration. This result is due to the second-derivative performed upon the Gaussian impulse. Additional derivatives taken on the Gaussian impulse would produce other pulses having the same pulse duration but with progressively increasing fractional bandwidth and frequency corresponding to the peak pulse-magnitude. This phenomenon further implies that UWB impulse signals generated using higher derivatives of the Gaussian impulse may be attractive for high-frequency UWB impulse systems since they have higher frequencies and larger fractional bandwidth for the same pulse duration, which may be useful for some applications. It is noted that using a Gaussian monocycle pulse, which is the first derivative of a Gaussian impulse, at high frequencies requires very narrow pulse duration, which may be difficult to realize with sufficient amplitude in practice.

As can be seen from the pulse waveforms, the Gaussian impulse has no zero crossing point, while the Gaussian monocycle pulse and Gaussian doublet pulse have one and two zero crossings, respectively, which help define the bandwidth characteristics of these pulses. It is also observed that the spectral responses of these pulses contain no side-lobes beyond the zero-crossing frequency points which are desirable for signal transmission. For pulses whose spectral responses have side-lobes, such as a rectangular or sinusoidal pulse, these side-lobes are always outside the pass-band, which at most extends across the zero-crossing frequency ends, and hence produce unwanted radiation, leading to possible false-target detection and/or interference to other existing systems, especially when they have sufficiently high energy.

It is particularly noted that, as the peak spectral amplitude of the Gaussian impulse occurs at dc and as seen in Fig. 2.3, the bulk of its energy is contained at dc and low frequencies near dc. The monocycle and doublet pulse signals, on the other hand, contain no dc component and have much lower low-frequency energy. In general, the monocycle and doublet pulses have similar energy distributions in the low- and high-frequency regions around the center frequency. It is the difference in the spectral shapes of these signals at dc and low frequencies that greatly affects the transmission of signals via antennas and the propagation of signals though components, and ultimately the design of UWB antennas, components and systems. Impulses are not transmitted and received effectively through practical antennas due to their large portion of low-frequency spectral components which cannot be transmitted (or is transmitted with very low efficiency) by practical antennas. Monocycle and doublet pulses, on the other hand, can be transmitted more efficiently due to no dc component and less low-frequency content. Furthermore, using monocycle or doublet pulse facilitates the design of components including antenna in UWB impulse systems due to no design consideration at dc and less design emphasis at low frequencies, leading to simpler and more compact design. It is further noted that signal fidelity is of utmost important for UWB impulse systems, which require signals to be transmitted and received with minimum distortion. With no dc component and less low-frequency spectral amplitudes contained in mono-cycle pulses, antennas and other system components can be more conveniently designed to cover desired bandwidth, hence minimizing the distortion of signals traveling through these components and, consequently, producing better fidelity for transmitting and receiving signals.

UWB impulse systems always transmit a train of pulses (typically periodically) instead of a single pulse. Consequently, according to Fourier analysis, the spectrums of UWB impulse signals are not continuous and contain discrete spectral lines (corresponding to discrete frequencies) spaced apart by 1/T, where T is the period of the UWB signals. Fourier analysis also shows that a UWB impulse signal consisting of a train of pulses is not substantially distorted by passive components including antennas having a bandwidth approximately equal to the reciprocal of the pulse width, because of most of the energy is contained within such bandwidth. According to the Parseval's theorem, the average power in a periodic pulse train is equal to the sum of the powers in its spectral components including dc and harmonics. Therefore, transmission of a UWB impulse signal consisting of periodic high-voltage pulses would be similar to simultaneous transmission of strong CW signals at different frequencies. The results of the Parseval's theorem also suggest an alternate way of generating a UWB impulse signal of periodic pulses by combining various CW signals having appropriate amplitudes and frequencies.

2.4   Basic Modulation Topologies

Several modulation techniques can be used to create modulated UWB signals, which modulate the information bits directly into very short UWB impulses [1]. Since there is no intermediate frequency (IF) processing in systems employing such signals, these systems are often called "base-band" or "impulse radio systems." Typical modulations in UWB impulse systems can be divided into the mono-phase and bi-phase techniques. The three most popular mono-phase UWB modulation approaches are the pulse position modulation (PPM), pulse amplitude modulation (PAM), and on-off keying (OOK). In these techniques, the data signal "1" is differentiated from the data "0" either by the size of the pulse signal or the time when it arrives with all the pulses essentially having the same shape. For the more efficient bi-phase case, the bi-phase shift keying (BPSK) is one of the most popular topologies. This modulation transmits a single bit of data with each pulse, with the positive pulse representing "1" and the negative pulse signifying "0". A brief description for each of these modulation topologies is gen as follows.

2.4.1   PPM

PPM is one of the common modulation technologies used in UWB impulse systems. In this technique, both the pulses that indicate digital data bits "1" and "0" have the same amplitude but at different times. The system transmits the same pulse at different positions in the time domain to represent a "1" or "0". This method may require a complex receiver in order to determine the precise position of the received pulse in order to recognize corresponding 1 or 0. An example of the PPM is shown in Fig. 2.8, where a (time) position of the pulse representing "1" leads that of the pulse representing "0".

Fig. 2.8 An example of PPM with pulses representing 1 and 0

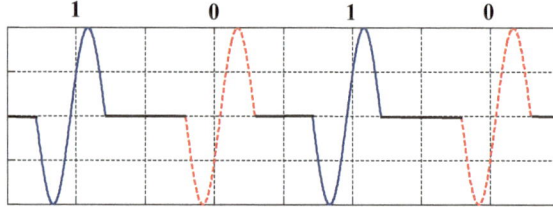

Fig. 2.9 An example of PAM with high- and low-amplitude pulse representing respective 1 and 0

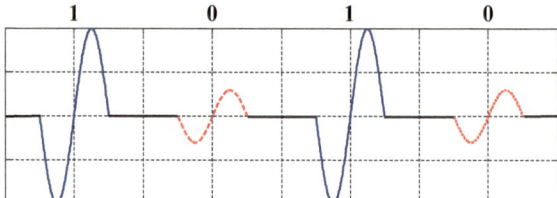

2.4.2 PAM

PAM works by separating the high-amplitude and the low-amplitude pulses. In the PAM, the amplitude of the pulse is varied according to different digital data information, where high-amplitude pulse represents "1" and small-amplitude pulse designates "0". Figure 2.9 illustrates an example of the PAM technique.

2.4.3 OOK Modulation

In OOK modulation, information bits "1" and "0" are represented with the full-amplitude and zero-amplitude of the UWB pulse, respectively, which are obtained by turning the UWB pulse on and off, respectively. By setting the UWB pulse on and off, binary information bits "1" and "0" are sent out. An example of the OOK modulation is shown in Fig. 2.10, where "1" is obtained when there is a pulse and "0" corresponds to no pulse.

Fig. 2.10 An example of OOK modulation where 1 and 0 are represented with and without pulse, respectively

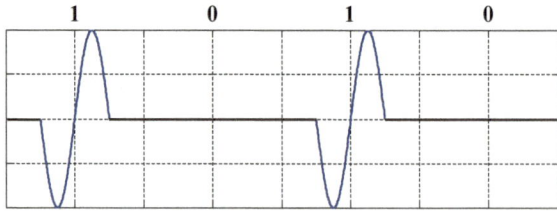

Fig. 2.11 An example of BPSK modulation with 1 and 0 represented by pulses of opposite polarities

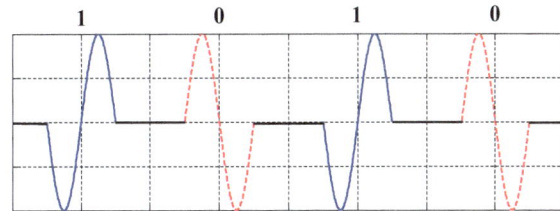

2.4.4 BPSK Modulation

The most common bi-phase modulation approach used in impulse UWB systems is BPSK. Bi-phase modulation differentiates "1" with a pulse (e.g., positive pulse) and "0" with another pulse of opposite polarity (e.g., a negative pulse). Comparing with the PPM, where a series of ultra-wideband circuits is needed to generate very accurate time steps, the bi-phase modulation approach is simple, requires only two kinds of pulses to be generated, and imposes less processing at the receiver side. The BPSK modulation offers several advantages such as power efficiency and smooth spectrum with smaller-amplitude spikes over the above-mentioned mono-phase techniques of PPM, OOK, and PAM that have larger amplitude spikes. These spikes are caused by the multi-pulse occurring periodically. The most significant advantage is an improvement of two times in the overall power efficiency as compared to the OOK or PPM [2]. This makes the bi-phase UWB approach extremely efficient for high-data-rate portable applications. An example of the BPSK modulation is shown in Fig. 2.11.

2.5 UWB Impulse Transmitters and Receivers

2.5.1 UWB Impulse Transmitters

One of the major advantages of impulse UWB transmitters, as compared to CW transmitters, is the simplicity of their circuits, in which complex components typically employed in CW transmitters, such as frequency synthesizers that contain various circuits like phase-locked-loop (PLL), voltage-controlled oscillator (VCO) and mixers, are not needed. Impulse UWB transmitters are thus relatively easier to design and implement and less expensive.

Figure 2.12 shows a block diagram of an UWB impulse transmitter, which is relatively simple and does not contain many components as seen in typical CW

Fig. 2.12 Block diagram of an impulse UWB transmitter

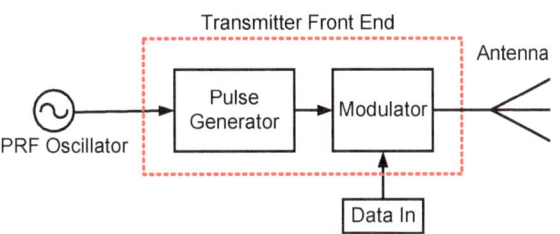

transmitters. It consists of a pulse generator and a digital-controlled modulator circuit that controls the timing or polarization of the transmitted pulse signal corresponding to digital data bit 1 or 0. The local oscillator, which typically employs a simple and inexpensive topology such as a crystal oscillator, determines the pulse repetition frequency (PRF) of the UWB impulse signal (and hence the corresponding UWB impulse system). The pulse generator produces a pulse signal with a desired waveform such as impulse or monocycle pulse, etc. The modulator circuit modulates the transmitting pulse signal with incoming digital data information using a particular modulation such as the BPSK, PAM, OOK or PPM described in Sect. 2.4, depending on the timing or polarization modulation requirement for the UWB impulse system.

The main component of the UWB impulse transmitters is the pulse generator. There are various ways to generate narrow-pulse signals in the pulse generator. Existing methods for generating sub-nanosecond pulses are generally based on hybrid circuits using discrete components, therefore resulting in relatively large size and high cost. These existing pulse generators are normally not optimized for minimum power consumption and feasibility of integrating them into wireless portable devices. Some of the pulse generators were developed using spark gaps [3], which are not an option for consumer electronics due to their size. One method of generating sub-nanosecond impulse and monocycle pulses involves hybrid circuits using Schottky barrier diodes and step recovery diodes (SRD) [4–6], which are also not very suitable for RFIC applications.

A periodic impulse can also be generated indirectly from individual concurrent sinusoidal signals having appropriate amplitudes and phases at different frequencies according to the Fourier series. A UWB impulse system employing such an impulse generator can be considered as equivalent to a frequency-domain system. Generation of impulse waveforms based on Fourier series is very accurate in theory [7, 8]. This technique, however, is very difficult to be implemented in practice at microwave frequencies due to the difficulty in generating and transmitting individual sinusoidal waveforms at different frequencies with precise amplitudes and phases as well as receiving them, especially when a large number of harmonics is needed to obtain an accurate waveform. A transmitter that can generate many harmonic components with certain phases and amplitudes is very difficult to design and could be bulky. In addition, the receiver design is also very complicated because the receiver needs to receive all of the returned signals from the transmitted harmonics for accurate results. Due to the complex nature of the transmitter and receiver design implementing this approach, it is not suitable for most wireless UWB applications, especially those requiring miniature and low-cost UWB impulse systems.

All of the foregoing mentioned impulse-generator designs have two common drawbacks: relatively large circuit size and possibly high cost (at least for mass production). These problems make them not desirable for compact UWB applications involving small-space and low-cost deployment. Using CMOS RFICs could resolve these issues because of their miniaturization, low cost, low power consumption, and easy integration with digital circuits (for complete single-chip systems).

Detailed design of UWB impulse transmitters is described in Chap. 3. Particularly, a RFIC pulse generator based on a commercial CMOS technology that produces both Gaussian impulse and monocycle pulse with tunable durations is presented. This pulse generator is integrated together with a BPSK modulator to form a UWB impulse transmitter module.

2.5.2   UWB Impulse Receivers

As for UWB impulse transmitters, one of the major advantages of UWB impulse receivers, as compared to CW receivers, is the simplicity of their circuits, in which mixers with intermediate-frequency (IF) stages typically employed in CW receivers are not needed. UWB impulse receivers directly convert received RF signals into a baseband output signal without an intermediate stage and are thus relatively easier to design and implement and less inexpensive.

Figure 2.13 shows a block diagram of a UWB impulse receiver, which consists of a low-noise amplifier (LNA), a correlator (correlation circuit), and a (template) pulse generator. The oscillator drives the pulse generator and determines the pulse repetition frequency (PRF) of the UWB impulse system. To maximize the processing gain and signal-to-noise ratio (SNR), the template waveform generated by the pulse generator should have a similar shape to that of the received pulse signal. After passing the LNA, the received pulse signal is coherently correlated with the template pulse through the correlator and the input pulse is then converted into a baseband signal. The conversion process can be done in a single stage (correlator) and, hence, no intermediate-frequency-conversion stage is needed, which greatly reduces the system complexity.

The correlator is the heart of UWB impulse receivers. It consists of a multiplier, an integrator, and a sampling/holding (S/H) circuit. The multiplier mixes or, precisely, multiplies the received pulse signal from the LNA with the (template) pulse signal from the pulse generator. The result (output signal) of the multiplier is integrated over several periods of the received pulse train to maximize the power and minimize the noise of the received signal. Due to the integration performed over a train of pulses, more correlated energy is integrated over the duration of each

Fig. 2.13 A UWB impulse receiver block diagram

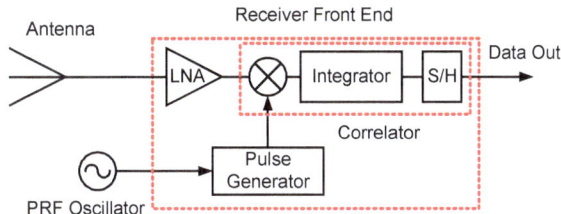

symbol (or received signal corresponding to the symbol) and hence the correlated signal is raised from the noise. Consequently, if more pulses (i.e., a longer pulse train) are used to transmit each symbol (essentially, the transmitting signal of the transmitter) or, in turn, contained in each symbol (essentially, the received signal of the receiver), then better SNR will be obtained.

Considering the unique feature of UWB impulse receivers described above, there is a stringent requirement for the correlation speed, which demands that both the multiplier and integrator must be fast enough to process each pulse. This inevitably brings great challenge to the correlator design for UWB impulse receivers.

Like most spread-spectrum systems, where energy generated in a particular bandwidth is deliberately spread over the operating spectrum in the frequency domain, resulting in a signal with a wider bandwidth, the processing gain (PG) is also an important characteristic in UWB impulse systems. To combat the unavoidable noise in and interference to the signal, a group or train of N pulses is used to transmit each symbol (signal), hence the energy of the symbol is spread over N pulses and the processing gain can be achieved. The processing gain (in dB) derived from this procedure can be defined as

$$PG_1 = 10 \, \log_{10}(N) \tag{2.13}$$

Furthermore, the pulse signal only occupies a very small part of the entire period, which means the duty cycle of the pulses can be extremely small, sometimes even less than 1%. Therefore, the UWB impulse receiver is only required to work for a small fraction of the period between the pulses, and the impact of any continuous source of interference is hence reduced so that it is only relevant when the receiver attempts to detect a pulse. The processing gain due to a low duty cycle is given by

$$PG_2 = 10 \, \log_{10}\left(\frac{T_f}{T_p}\right) \tag{2.14}$$

where T_f is the period and T_p is the pulse width.

The total processing gain PG is the sum of the two processing gains:

$$PG = PG_1 + PG_2 = 10 \, \log_{10}(N) + 10 \, \log_{10}\left(\frac{T_f}{T_p}\right) \tag{2.15}$$

As an example, consider a UWB impulse receiver operating with a pulse train having a pulse period of 100 ns and a pulse width of 200 ps, the processing gain due to the low duty cycle (PG_2) would be about 27 dB. Since the UWB impulse system uses multiple pulses to recover each bit of information (symbol), if one digital bit is determined by integrating over 100 pulses, then the processing gain PG_1 would be another 20 dB. The total processing gain PG for the UWB impulse system is then about 47 dB. Since the PRF of the pulse is 10 MHz and each bit covers 100 pulses, the resulting data rate is 100 Kbps as obtained from (2.15).

Detailed design of UWB impulse receivers is given in Chap. 4. Particularly, the designs of a UWB LNA, correlator and receiver on RFIC using a commercially available CMOS technology are described.

2.6 UWB Antennas

UWB antennas capable of radiating and receiving faithfully UWB impulse signals are crucial for high-performance UWB impulse systems. As all of the frequency components contained in an impulse signal needs to be transmitted or received concurrently, UWB antennas used in UWB impulse systems have more stringent requirements than UWB antennas employed in UWB CW systems, which only need to transmit or receive one frequency component at each time. These strict requirements, such as minimum dispersion and loss (and hence distortion), make the design of UWB impulse antennas more difficult, particularly on planar circuits that enable direct and low-cost integration with UWB impulse transmitters and receivers. Chapter 5 discusses UWB antennas for UWB impulse systems and presents the detailed design of a uniplanar UWB antenna implemented on microwave integrated circuits, that is suitable for UWB impulse systems.

2.7 Summary

This chapter covers the fundamentals of UWB impulse systems operating across or within the unlicensed UWB frequency band of 3.1–10.6 GHz. It provides the essence of UWB impulse systems including the spectrum mask, advantages and applications of UWB impulse systems, UWB impulse signals including Gaussian impulse, doublet pulse and monocycle pulse, modulations including PPM, PAM, OOK and BPSK, UWB impulse transmitters and receivers, and UWB antennas.

References

1. H. Arslan, Z.N. Chen, M.G. Di Benedetto, *Ultra Wideband Wireless Communications* (John Wiley & Sons Inc, Hoboken, NJ, 2006)
2. V.G. Shpak, M.R. Oulmascoulov, S.A. Shunailov, M. I. Yalandin, Active former of monocycle high-voltage subnanosecond pulses. 12th IEEE Int. Pulse Power Conf. **2**, 1456–1459 (1999)
3. J.W. Han, C. Nguyen, On the development of a compact sub-nanosecond tunable monocycle pulse transmitter for UWB applications. IEEE Trans. Microwave Theory Tech. **MTT-54**(1), 285–293 (2006)

4. J.S. Lee, C. Nguyen, Novel low-cost ultra-wideband, ultra-short-pulse transmitter with MESFET impulse-shaping circuitry for reduced distortion and improved pulse repetition rate. IEEE Microwave Wirel. Compon. Lett. **11**(5), 208–210 (2001)
5. J. Han, C. Nguyen, Ultra-wideband electronically tunable pulse generators. IEEE Microwave Wirel. Compon. Lett. **14**(3), 112–114 (2004)
6. G.S. Gill, H.F. Chiang, J. Hall, Waveform synthesis for ultra wideband radar, *Record of the 1994 IEEE National Radar Conference*, pp. 29–31, Mar 1994
7. G.S. Gill, Ultra-wideband radar using Fourier synthesized waveforms. IEEE Trans. Electromagn. Compat. **39**(2), 124–131 (1997)
8. I. Oppermann, M. Hamalainen, J. Iinatti, *UWB Theory and Applications* (John Wiley & Sons Inc, Hoboken, NJ, 2004)

Chapter 3
UWB Impulse Transmitter Design

3.1 Introduction

Typical UWB impulse transmitters, as introduced in Sect. 2.5.1, consist of two fundamental components: pulse generator and modulator. Pulse generator is also a key component providing the template pulse for UWB impulse receivers as mentioned in Sect. 2.5.2. In essence, in view of the simplicity of UWB impulse transmitters, we can somewhat consider a UWB impulse transmitter as a pulse generator, making its topology and design much simpler than those of a CW transmitter. Depending on the requirements of a particular UWB impulse system, the pulse generator can be designed to generate different pulse signals—among them; the Gaussian-like impulse and monocycle pulse are typical as discussed in Sect. 2.3. Moreover, the generated UWB impulse signal should also meet other specifications of a UWB impulse system. The modulator, depending on applications, can employ different modulation schemes such as PPM or BPSK as addressed in Sect. 2.4 to modulate the information bits directly into the short pulses transmitted by a UWB impulse transmitter.

The most essential specifications of the pulse generator are the duration and peak power of the output pulse. The duration of the output pulse should be sufficiently narrow to provide the required resolution for accurate detection of targets or ranges. The peak power of the output pulse needs to be high enough to enable detection of targets at long ranges. The generated pulse signal should also have good shape and symmetrical with minimum distortion in the main pulse and minimum ringing-tails (or side-lobes) of the main pulse. Another desirable function of pulse generators is the pulse-duration tuning for generating pulse signals of different durations, preferably with near-constant peak power and pulse shape.

Tunable pulse generator can provide great flexibility in the operation of UWB impulse systems with enhanced performance adaptable to different environments and targets [1, 2]. Pulse with wide duration contains large low-frequency components, enabling the pulse signal to propagate farther because of the relatively low

C. Nguyen and M. Miao, *Design of CMOS RFIC Ultra-Wideband Impulse Transmitters and Receivers*, SpringerBriefs in Electrical and Computer Engineering, DOI 10.1007/978-3-319-53107-6_3

propagation loss of its low-frequency components. Pulse with shorter duration, on the other hand, has wider frequency bandwidth, making feasible higher range resolution. A pulse that can change its duration, especially by an electronic means, would therefore have both advantages of increased range (or penetration) and fine range resolution, and is attractive for UWB impulse systems, especially those intended for sensing of wide variety of targets that need varying ranges and resolutions. The pulse tune-ability also provides the highly desired diversity for UWB impulse transmitters (and hence UWB systems), enabling them to function across multi-band which leads to enhanced information of targets. Electronically tunable pulse generators are also desired for measurement equipment. Tuning ability is also useful for compensating possible circuit performance variations caused by potential variations of circuit fabrication processes such as CMOS processes as well as temperature changes. UWB tunable impulse and monocycle pulse transmitters were developed using step-recovery and PIN diodes and hybrid circuits [2, 3].

Most existing UWB pulse generators were designed using microwave integrated circuits (MIC), which are relatively large and expensive (for mass production to some extent). As discussed in Chap. 1, miniature and low-cost UWB impulse systems (and hence UWB impulse transmitters) are desired for small size and low cost, making the CMOS RFIC attractive for UWB impulse transmitters, especially for portable or handheld commercial UWB systems developed for wireless communications and sensors.

Several CMOS RFIC pulse generator topologies were reported for UWB communications using IBM 0.18-μm BiCMOS [4], TSMC 0.18-μm CMOS [5, 6], and CSM 0.18-μm CMOS/BiCMOS [7] processes. However, no experimental results were presented for these circuits. The calculated pulse-widths and amplitudes of the output pulses are around 300 ps and 20 mV [8], 300 ps and 22.97 mV [5], 380 ps and 650 mV [6], and 200 ps and 27 mV [7]. Furthermore, a square-wave and a 1.5-GHz clock signal were used externally as the input in [4, 7] and [5] for simulations, respectively. A CMOS RFIC tuneable pulse generator on TSMC 0.25-μm CMOS process that can generate monocycle pulse and Gaussian-like impulse signals was also designed, fabricated, and measured for UWB impulse systems [9].

In this chapter, we present the design of two UWB impulse transmitters having sub-nanosecond pulses with tunable pulse-width [10]: a UWB impulse transmitter and a UWB monocycle-pulse transmitter. The CMOS UWB impulse transmitter integrates a pulse generator and a BPSK modulator together, and the CMOS UWB monocycle-pulse transmitter consists of the CMOS UWB impulse transmitter and a pulse-forming circuitry. The chapter also includes the design of CMOS impulse and monocycle pulse generators. The CMOS impulse generator generates 0.95–1.05 V peak-to-peak Gaussian-type impulse signal with 100–300 ps tunable pulse duration. The CMOS monocycle pulse generator produces 0.7–0.75 V peak-to-peak monocycle pulse with 140–350 ps tunable pulse duration. These pulse signals can be used for various UWB impulse systems. An external clock signal operating at a low frequency of only 10 MHz is used. The BPSK modulator controls the pulse generator to generate positive or negative pulse signal depending on the "1" or "0" digital data information.

This chapter is arranged as follows. Firstly, some basic components used in the CMOS UWB impulse transmitter design, such as CMOS inverter, delay cell, and NOR/NAND gate block, are briefly described. Secondly, the designs of the CMOS UWB impulse transmitters' constituent tunable pulse generators and BPSK modulator are covered. Finally, the CMOS UWB impulse transmitters realized by integrating the fully integrated tunable pulse generators with the BPSK modulator are presented.

3.2 Overview of Basic Components of UWB Impulse Transmitters

3.2.1 CMOS Inverter Basics

CMOS inverter is an essential element in the tunable pulse generator design constituted by delay cell and square-wave generator and plays an important role in the operation of these constituents. In fact, the CMOS inverter is a basic building block for digital circuit design. However, the operation of the CMOS inverter in tunable pulse generators is somewhat different with that of the digital case. As shown in Fig. 3.1, the CMOS inverter consists of a pair of enhancement-type NMOS and PMOS transistors operating in complementary mode. The input voltage V_{in} is fed to the gates of both transistors. The substrate (or body or bulk) of the NMOS transistor is connected to the ground, while the substrate of the PMOS transistor is connected to the power supply voltage V_{dd}, to reverse-bias the source and drain junctions. Since the voltage between the source and body $V_{SB} = 0$ for both transistors, there is no substrate-bias effects.

Comparing with other inverter configurations, such as the MOS current mode logic (MCML), the CMOS inverter has two important advantages. The first and most important one is its virtually negligible steady-state power dissipation, except for small power dissipation due to leakage currents. On the other hand, in other inverter structures like MCML, a nonzero constant steady-state current is drawn from the power source when the driver transistor is turned on, which results in a

Fig. 3.1 CMOS inverter circuit and its symbol

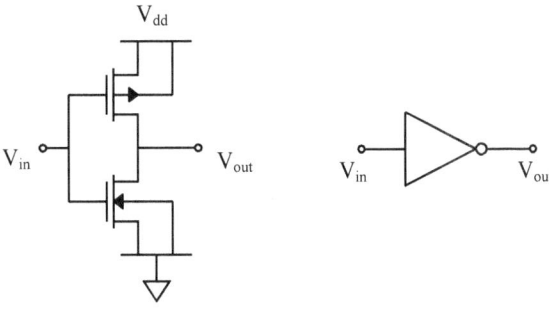

significant dc power consumption. The other advantage of the CMOS inverter configuration is that the voltage transfer characteristic (VTC) exhibits a full output voltage swing between 0 V and V_{dd}. On the contrary, the MCML has a much lower swing voltage than V_{dd}, which cannot provide enough driving voltage and hence is not suitable to the tunable pulse generator design.

The inverter's threshold voltage V_{th}, which is considered as the transition voltage and defined as the point where $V_{in} = V_{out}$, is an important parameter characterizing the steady-state input-output behavior of the CMOS inverter [11]. This voltage is given by

$$V_{th} = \frac{V_{T0,n} + \sqrt{\frac{1}{k_R}} \cdot (V_{dd} + V_{T0,p})}{\left(1 + \sqrt{\frac{1}{k_R}}\right)} \tag{3.1}$$

where V_{dd} is the power supply, $V_{T0,n}$ is the NMOS threshold voltage, $V_{T0,p}$ is the PMOS threshold voltage, and k_R is defined as

$$k_R = \frac{k_n}{k_p} \tag{3.2}$$

k_n and k_p are the transconductance parameters given as

$$k_n = \mu_n \cdot C_{ox} \cdot \frac{W}{L} \tag{3.3a}$$

$$k_p = \mu_p \cdot C_{ox} \cdot \frac{W}{L} \tag{3.3b}$$

where μ_n is the electron mobility in the NMOS transistor, μ_p is the hole mobility in the PMOS transistor, C_{ox} is the gate-oxide capacitance, and W and L are the channel width and length, respectively.

Since the CMOS inverter is a fully complementary structure, to achieve a completely symmetric output signal, the threshold voltages are set as $V_{T0} = V_{T0,n} = |V_{T0,p}|$. Therefore [7],

$$\left(\frac{k_n}{k_p}\right)_{\substack{symmetric \\ inverter}} = 1 \tag{3.4}$$

We can obtain from (3.3):

$$\frac{k_n}{k_p} = \frac{\mu_n C_{ox} \cdot \left(\frac{W}{L}\right)_n}{\mu_p C_{ox} \cdot \left(\frac{W}{L}\right)_p} = \frac{\mu_n \cdot \left(\frac{W}{L}\right)_n}{\mu_p \cdot \left(\frac{W}{L}\right)_p} \tag{3.5}$$

where $(W/L)_n$ and $(W/L)_p$ correspond to the NMOS and PMOS transistors, respectively. The unity-ratio condition in (3.4) for the ideal symmetric inverter requires that

$$\frac{\left(\frac{W}{L}\right)_n}{\left(\frac{W}{L}\right)_p} = \frac{\mu_p}{\mu_n} \tag{3.6}$$

It is noted that μ_p is typically much smaller than μ_n, making $(W/L)_n \ll (W/L)_p$. Consider $\mu_p = 203 \text{ cm}^2/\text{V s}$ and $\mu_n = 580 \text{ cm}^2/\text{V s}$ [7], to achieve the symmetric input-output performance in the CMOS inverter design, we approximately select the dimension ratio of the PMOS transistor to NMOS transistor as

$$\frac{\left(\frac{W}{L}\right)_p}{\left(\frac{W}{L}\right)_n} = 3 \tag{3.7}$$

Assume the same minimum gate length L is used for both PMOS and NMOS transistors, we then obtain $W_p = 3W_n$.

3.2.2   CMOS Inverter Switching Characteristics

We begin by introducing some commonly used delay times as defined in Fig. 3.2 including the propagation delay times τ_{PHL}, τ_{PLH} and the rising τ_{rise} and falling τ_{fall} times. As illustrated in Fig. 3.2, the delay times τ_{PHL} and τ_{PLH} determine the input-to-output signal delay during the high-to-low and low-to-high transitions of the output signal, respectively. Specifically, τ_{PHL} is the time delay between the transition at 50% $(V_{50\%})$ of the rising input voltage (top figure) and the transition at 50% $(V_{50\%})$ of the falling output voltage (middle figure). Similarly, τ_{PLH} is defined as the time delay between the transition at 50% $(V_{50\%})$ of the falling input voltage (top figure) and the transition at 50% $(V_{50\%})$ of the rising output voltage (middle figure) [7]. As seen in Fig. 3.2, the rising time τ_{rise} is defined as the time between 10 and 90% of the rising output voltage; that is, the time needed for the output voltage to rise from the $V_{10\%}$ level to the $V_{90\%}$ level (bottom figure). Similarly, the falling time τ_{fall} is defined as the time between 90 and 10% of the falling output voltage or the time for the output voltage to drop from the $V_{90\%}$ level to the $V_{10\%}$ level [11].

Assuming the input signal is a step pulse with zero rise and fall times, the propagation delay time for the high-to-low output transition τ_{PHL} of a CMOS inverter can be expressed as [11]

$$\tau_{PHL} = \frac{C_{load}}{k_n\left(V_{dd} - V_{T,n}\right)}\left[\frac{2V_{T,n}}{V_{dd} - V_{T,n}} + \ln\left(\frac{4\left(V_{dd} - V_{T,n}\right)}{V_{dd}} - 1\right)\right] \tag{3.8a}$$

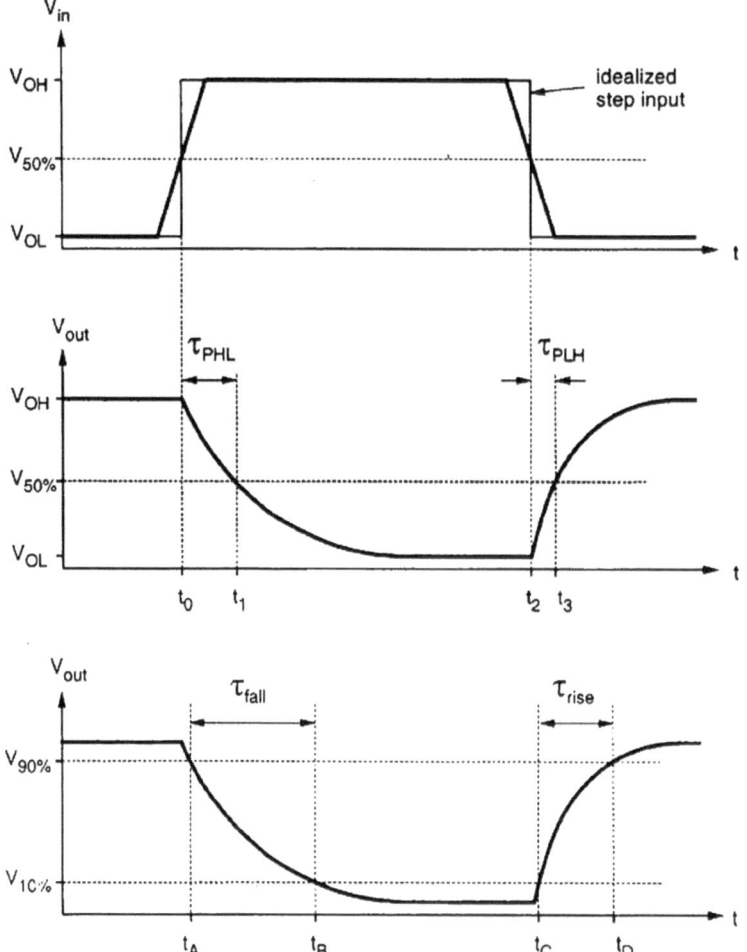

Fig. 3.2 CMOS inverter delay-time definitions

Similarly, the propagation delay time from the low-to-high output transition τ_{PLH} of a CMOS inverter is [11]

$$\tau_{PLH} = \frac{C_{load}}{k_p\left(V_{dd} - |V_{T,p}|\right)} \left[\frac{2|V_{T,p}|}{V_{dd} - |V_{T,p}|} + \ln\left(\frac{4\left(V_{dd} - |V_{T,p}|\right)}{V_{dd}} - 1\right) \right] \quad (3.8b)$$

where C_{load} is the load capacitance at the output node. Equations (3.8a) and (3.8b) show that, in order for $\tau_{PHL} = \tau_{PLH}$, the conditions $V_{T,n} = |V_{T,p}|$ and $k_n = k_p$ must be satisfied.

Considering a situation where the input voltage waveform is not an ideal pulse waveform, but has finite rising (τ_r) and falling (τ_f) times, the corresponding propagation delay times can be empirically expressed as [7]

$$\tau_{PHL}(actual) = \sqrt{\tau_{PHL}^2(step\ input) + \left(\frac{\tau_r}{2}\right)^2} \qquad (3.9a)$$

$$\tau_{PLH}(actual) = \sqrt{\tau_{PLH}^2(step\ input) + \left(\frac{\tau_f}{2}\right)^2} \qquad (3.9b)$$

where $\tau_{PHL}(step\ input)$ and $\tau_{PLH}(step\ input)$ are the propagation delay times for the step pulse input waveform given in (3.8).

Assuming the input voltage of the CMOS inverter is an ideal step waveform with negligible rising and falling times, the average power dissipation of the CMOS inverter can be written as [11]

$$P_{avg} = C_{load} \cdot V_{dd}^2 \cdot f \qquad (3.10)$$

where f is the switching frequency, which is the frequency of the clock signal. From (3.10) we can find that the average power dissipation of the CMOS inverter is proportional to the switching frequency f. For the pulse generator design presented in this chapter, the clock frequency is only 10 MHz. Therefore, comparing with the MCML structure, where the static power consumption is $V_{dd} \cdot I$, the CMOS inverter has a much smaller average power dissipation.

3.2.3   Two-Input CMOS NOR/NAND Gate Blocks

Figure 3.3 shows the circuit schematic and symbol of a two-input CMOS NOR gate block. The circuit consists of series-connected complementary PMOS transistors and parallel-connected NMOS transistors. The input voltages V_A and V_B are applied to the gates of NMOS and PMOS transistors.

For the NOR gate block, the output voltage is high only under the condition that both input signals V_A and V_B are low voltages. For the other conditions, the output voltage is always low. Based on this characteristic, the NOR gate block can be used to generate positive impulse signal in the pulse generator design by adjusting the time difference between the two low-voltage input signals to a very small value.

Figure 3.4 shows the circuit schematic and symbol of a two-input CMOS NAND gate block. The circuit consists of parallel-connected PMOS transistors and series-connected complementary NMOS transistors. The operating principle is exact dual of the CMOS NOR gate block. Therefore for the NAND gate block, the output voltage is low only at the condition that both V_A and V_B are high voltages. For all other

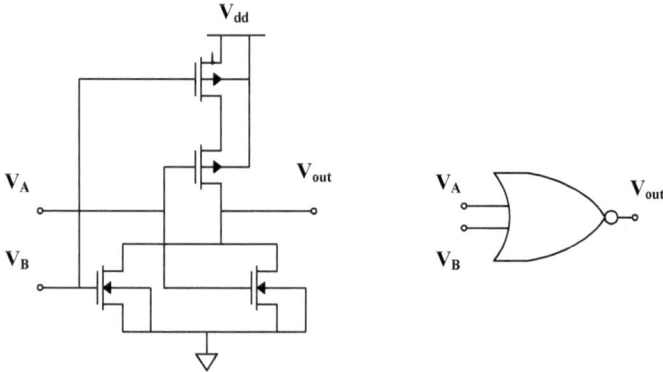

Fig. 3.3 CMOS two-input NOR (NOR2) gate block and its symbol

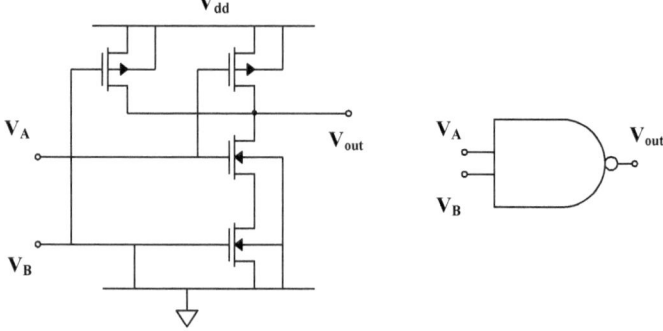

Fig. 3.4 CMOS two-input NAND (NAND2) gate block and its symbol

conditions, the output voltage is always high. Hence, the NAND gate block can be used to generate negative impulse signal in the pulse generator design by controlling the time difference between the two high-voltage input signals to a small value.

3.2.4   Tunable Delay Cell

Variable delay elements are often used to manipulate the rising or falling edges of the clock signal or any other signals in ICs. There are three different kinds of delay-element architectures in CMOS VLSI design: transmission-gate based, cascaded-inverter based, and voltage-controlled based [11]. Here we select the voltage-controlled shunt-capacitor delay element as the tuning delay cell in the tunable pulse generator design because of its relatively simple structure [12, 13].

Figure 3.5 shows the basic circuit of the voltage-controlled shunt-capacitor delay element. It consists of a shunt-controlled transistor M1 and the shunt MOS

Fig. 3.5 Shunt-capacitor
delay element

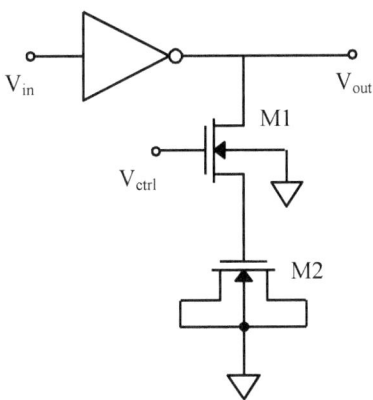

capacitor M2. The control voltage V_{ctrl} adjusts the resistance of the shunt transistor M1, which connects the load capacitance M2 to the output of a logic stage. The tuning voltage V_{ctrl} modulates the resistance of the shunt transistor M1, which is equivalent to changing the effective shunt capacitor value to the output of the inverter. Larger value of V_{ctrl} decreases the resistance of the shunt transistor M1, so the effective shunt capacitance at the logic gate output is bigger, producing a larger time delay. By selecting a suitable size for the shunt capacitor M2 with respect to a specific output capacitor load, a desired continuous time tuning range can be achieved. The tunable capability of this shunt-capacitor delay element plays an important role in the tunable pulse generator design, and the details will be described in the next section.

3.3 Tunable Pulse Generator Design

Pulse generator is the key component in UWB impulse systems. It can function as a source for the UWB impulse transmitter or as an internal source for the template signal in the UWB impulse receiver. Figure 3.6 shows the block diagram of the CMOS UWB tunable monocycle pulse generator. It integrates a tuning delay circuit, a square-wave generator, an impulse-forming circuit, and a pulse-shaping circuit in a single chip.

3.3.1 Tuning Delay Component

The tuning delay component includes a pair of parallel tunable delay cell and reference cell using shunt-capacitor delay elements [12], as shown in Fig. 3.7. M2 is a NMOS-type capacitor. The NMOS transistor M1 controls the charging and

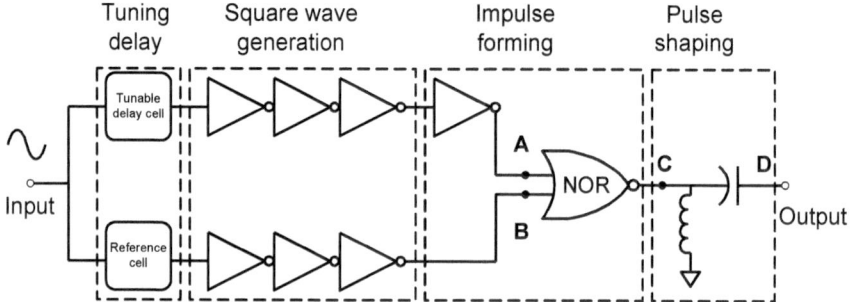

Fig. 3.6 Block diagram of the CMOS UWB tunable monocycle pulse generator

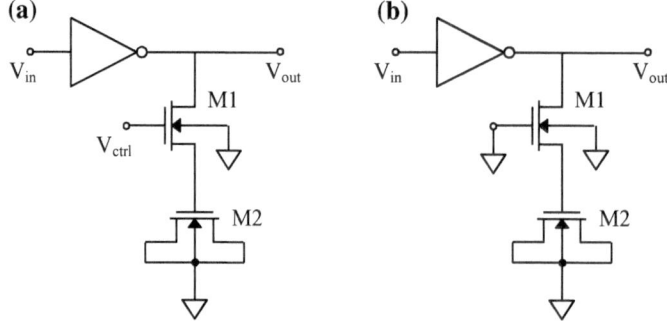

Fig. 3.7 Circuit schematics of the tunable delay cell (**a**) and reference cell (**b**)

discharging current to the capacitor M2. The only difference between the circuits of the tunable delay cell and reference cell is the gate voltage of the shunt transistor M1, which controls the charge current. For the tunable delay cell, variable control voltage V_{ctrl} between 0 V and V_{dd} is applied to the gate of the transistor M1 to produce continuous delay variation. On the other hand, for the reference cell, the gate of the transistor M1 is directly connected to the ground, so the gate voltage of M1 is fixed to zero, therefore the time-delay is constant and provides a reference position to the tunable delay cell.

There are several advantages of using two identical delay structures. One is that the relative time-delay between the two paths can be easily controlled as can be inferred from the schematic. Another advantage is enabling the generation of a pulse signal with extremely small pulse width as explained in the following. For single delay cell situation, assuming perfect condition, the delay cell is equivalent to an infinitesimal capacitor when the gate voltage of the shunt transistor M1 is 0 V. However, in reality, the delay cell is non-perfect, thereby there always exists leakage current inside the shunt transistor M1, effectively making the capacitor having a finite value. That means, for the method where only single delay cell is used, an inherent minimum absolute time delay, caused by the non-perfect delay cell, always exists, and the value of this time delay is sometimes much larger than the minimum pulse width required to achieve. Therefore, single delay cell topology impedes the design of pulse generators with extremely narrow pulse signal, and hence is not desired even it occupies less die area. With the adoption of the parallel delay elements, this parasitic time delay effect can be eliminated, and the minimum relative time difference achieved can be as small as possible. This advantage guarantees the pulse generator to produce a pulse signal with extremely small pulse width. The larger the value of the capacitor M2, the broader the tuning range of the pulse width. Hence, by tuning the gate-controlled voltage V_{ctrl} within the range from 0 V to V_{dd}, a pulse signal with different pulse widths can be achieved.

It is noted that the pair of parallel delay cells is located at the first stage of the entire pulse generator circuit, directly in front of the square wave generator. This arrangement helps reduce some strict requirements for the delay element design and hence facilitates its design. The input signal of the pulse generator is a sinusoidal signal with frequency of only 10 MHz, which is much lower than the maximum operating frequency of the CMOS inverter and delay cell. The extra capacitor load introduced by the tuning delay cell has negligible effects on the rising and falling times of the final output signal of the CMOS inverter with a reasonable tuning range of the time delay around 500 ps. If the tuning delay cells succeed the square wave generator and precede the impulse-forming circuit, which seems a straightforward topology to generate tunable delay at the first glance, it would actually bring other potential problems to the delay cell design. The reason is very clear, the signal produced by the square wave generator has very sharp rising and falling edges, normally in the order of less than 100 ps, which correspond to frequency components of more than 10 GHz. After passing the delay cell, the signal should keep the same rising and falling times, which means the operating frequency of the delay cells should be more than 10 GHz, which makes it difficult to design the delay cell. Furthermore, the input capacitor of the next-stage impulse-forming circuit is very large because of the driving capability requirement. To achieve the same time delay range of 500 ps, the size of the shunt-capacitor M2 should increase dramatically.

3.3.2 Square Wave Generator

For the tunable pulse generator design using two parallel signal paths as shown in Fig. 3.6, the tunable delay cells that can produce an extremely short time difference between the two signal paths are certainly important for generating a pulse signal with a very narrow pulse width. However, it is still not enough if only these components are involved in the pulse generation. There are other parameters affecting the final pulse signal performance, which are the rising and falling times of the generated square wave signal.

The function of the square-wave generator is to produce a square-wave signal with very short rising and falling times when a sinusoidal clock signal is fed to the circuit. Sharp rising/falling time is needed as the minimum width of the impulse

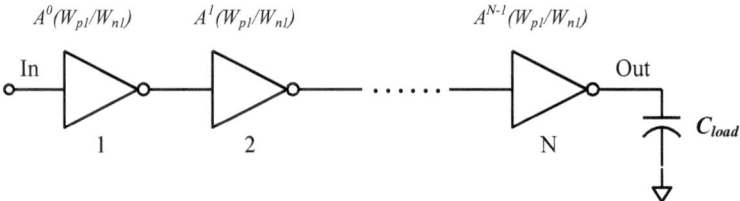

Fig. 3.8 Cascade of inverters used to drive a large load capacitor

signal generated in the subsequent stage is determined by the rising and falling times of the feeding square-wave. The succeeding stage of the square wave generator is the impulse-forming block, whose size should be large enough to provide the driving capability for the next stage circuit; therefore, the input capacitor of the impulse-forming block is very large. To drive this large capacitor effectively without sacrificing the rising/falling edge performance, a series of CMOS inverters with increasing size for each step (i.e., a buffer circuit) is used to increase the drive capabilities and shorten the rising and falling times of the square-wave signal.

Consider a circuit consisting of cascaded N inverters driving a load capacitor C_{load} as shown in Fig. 3.8, where A is a constant larger than 1, and W_{p1} and W_{n1} are the channel widths of the PMOS and MOS of the first CMOS inverter, respectively. Each inverter, as seen in Fig. 3.8, is A times larger than the previous one; therefore, each inverter's input capacitance is larger than the previous inverter's input capacitance by a factor of A as [14]

$$C_{in2} = A \cdot C_{in1}, C_{in3} = A^2 \cdot C_{in1}, \ldots, C_{inN} = A^{N-1} \cdot C_{in1} \tag{3.11}$$

The corresponding effective switching resistances are [13]

$$R_{n,p2} = \frac{R_{n,p1}}{A}, R_{n,p3} = \frac{R_{n,p1}}{A^2}, \ldots \ldots, R_{n,pN} = \frac{R_{n,p1}}{A^{N-1}} \tag{3.12}$$

where the effective switching resistance of the first CMOS inverter $R_{n,p1}$ is defined as [14]

$$R_{n,p1} = \frac{V_{dd}}{\frac{\mu_n C_{ox}}{2}(V_{dd} - V_{THN})^2} \cdot \frac{L}{W_{n1}} + \frac{V_{dd}}{\frac{\mu_p C_{ox}}{2}(V_{dd} - V_{THP})^2} \cdot \frac{L}{W_{p1}} \tag{3.13}$$

Therefore, for each stage of the buffer, the same delay of $R_{n,p1} \cdot C_{in1}$ is achieved.

Typically, it is assumed that the load capacitance C_{load} has the following relation with the input capacitance of the last inverter of the buffer [14]:

$$C_{load} = A \cdot C_{inN} = A^N \cdot C_{in1} \tag{3.14}$$

from which, we can determine the factor A as

$$A = \left(\frac{C_{load}}{C_{in1}}\right)^{1/N} \tag{3.15}$$

Therefore, the total time delay of the inverter buffer can be found from [14]

$$(\tau_{PHL} + \tau_{PLH})_{total} = 0.7N\left(R_{n1} + R_{p1}\right)\left(C_{out1} + AC_{in1}\right) \tag{3.16}$$

where C_{out1} is the total capacitance at the output of the first inverter, which includes the sum of the output capacitance of the inverter, any capacitance of interconnecting lines, and the input capacitance of the following stage.

For the square wave generator used in the tunable pulse generator design, the minimum time delay is not an essential parameter, so the value of the factor A can be selected other than 2.72 (corresponding to the minimum time delay condition) [14] to reduce the stage number of the square wave generator. During the square wave generator design, the size of the inverters sets the rising and falling times; however, there are some constraints that should be considered. One is the size of the first stage inverter should be chosen with a small value, so that the input capacitance C_{in1} of the CMOS inverter is small enough to maintain a large enough ratio of C_{shunt} to C_{in1}, where C_{shunt} is the capacitor M2 of the tuning delay cell. Therefore, a sufficiently broad tuning range can be achieved. If the size of the first stage inverter increases too much, to keep the same time-delay tuning range, the size of the shunt-capacitor in the tuning delay element has to increase accordingly, which will consume more die area. Another is that, if too many inverters are used, according to Eq. (3.12), the number of the buffer stages N in the square-wave generator must be increased, which increases the power consumption. Another requirement is that the rising and falling times should be approximately the same, which is important for symmetric pulse signal generation. This is accomplished by making the PMOS transistor about 3 times larger than the NMOS transistor in each inverter because of the different carrier mobility in PMOS and NMOS transistors [14].

3.3.3 Impulse-Forming Block

An impulse-forming block can be used to generate a positive or negative impulse signal, depending on which gate block is selected as the impulse-forming core. If the NOR gate block is used in the impulse-forming block, then the output signal is a positive impulse. On the other hands, if the NAND gate block is used, a negative impulse signal would be generated. Only the impulse-forming block with a positive impulse signal is described in this section; the analysis and design of the impulse-forming block with a negative impulse are similar.

The impulse-forming block is made up of an inverted delay stage and a NOR gate block, as shown in Fig. 3.6. The main purpose of the NOR gate block is to generate a positive impulse-like signal and provide driving capability to the next stage. This impulse should also be able to evoke the impulse response of the succeeding component to further produce a monocycle pulse (or other kind of pulse waveforms as needed for UWB impulse systems). The function of the inverted delay stage is to provide one input to the NOR gate block with a square wave signal, which is the reverse replica of the other input signal. As the signal produced by the square wave generator has extremely narrow symmetric rising and falling edges, the size of this inverter should be selected to provide enough driving capability to maintain the same rising and falling edges for the output signal. With the help of the previous-stage tuning delay component, the time difference between the two input signals of the NOR gate block can be adjusted continuously to generate a positive impulse signal with tunable pulse width.

3.3.4   Pulse-Shaping Circuit

The last stage of the tunable monocycle pulse generator is the pulse-shaping circuit, which consist of a shunt on-chip spiral inductor and a series metal-insulator-metal (MIM) capacitor operating as a high pass filter (HPF). The on-chip octagonal shape spiral inductor is designed and optimized using the EM software IE3D [15] to achieve improved quality factor Q instead of using the inductor provided in the foundry design kit. By optimizing the values of the spiral inductor and MIM capacitor, the pulse-shaping circuit functions approximately like a differentiator for the generated tunable impulse signal. As a result, a monocycle pulse signal with tunable duration can be generated when the impulse-like signal from the impulse-forming circuit is fed to the pulse-shaping circuit.

3.3.5   Tunable Pulse Illustration

Figure 3.9 illustrates the voltage variations at different nodes A, B, C, and D of the tunable monocycle pulse generator designated in Fig. 3.6 when a 10-MHz sinusoidal clock signal is fed to the generator.

As shown in Fig. 3.6, the input clock signal is divided equally into two paths: one signal passing through the tunable delay cell in the top path and another going through the reference cell in the bottom path. At node B, a square-wave signal (0 V to V_{dd}) with very short rising and falling times is generated and functions as one of the inputs to the following NOR gate block. By choosing a suitable control voltage V_{ctrl} between 0 V and V_{dd} for the tunable delay cell, another square wave with a different delay time is generated at node A. This signal is the reversed replica of that at node B with a certain time difference and acts as another input signal to the NOR

Fig. 3.9 Illustration of signal shapes at each node of the tunable pulse generator shown in Fig. 3.6

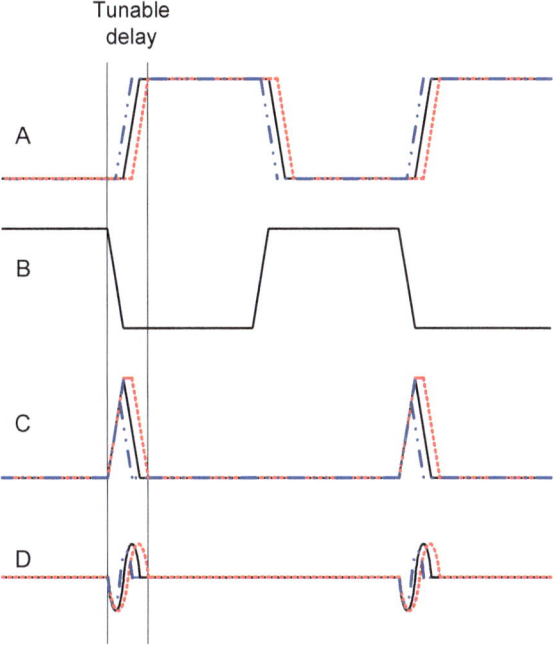

gate block. The output of the NOR gate block is at high state (V_{dd}) only when the inputs to the NOR gate are both at low state (0 V). For all the other input states, the output are always low (0 V). When these two reversed square waves at A and B are fed to the NOR gate block, a narrow impulse-like signal is generated at node C. The width of this impulse signal depends on the relative time difference between these two square-wave signals and the widths of their rising and falling edges. The impulse signal at node C, therefore, can be easily generated with a continuously tuning duration. A smaller time difference between nodes A and B generates a narrower impulse with a smaller peak-to-peak voltage on node C, while a larger time difference produces a broader impulse with a higher peak-to-peak voltage. When the tunable impulse signal is sent to the pulse-shaping circuit, a monocycle pulse signal with different durations is achieved at node D.

3.3.6 Simulation and Measurement Results

The tunable pulse generator was fabricated using Jazz 0.18-µm CMOS process [16]. The design and simulation were performed using the Agilent Advanced Design System (ADS) [17], Cadence Design Systems [18] and the Jazz 0.18-µm CMOS Process Design Kit (PDK). A single 1.8-V supply voltage was used for the entire circuit.

Fig. 3.10 Photograph of the
0.18-µm CMOS tunable
monocycle pulse generator
chip including pads for
on-wafer probe measurement

Tunable
delay

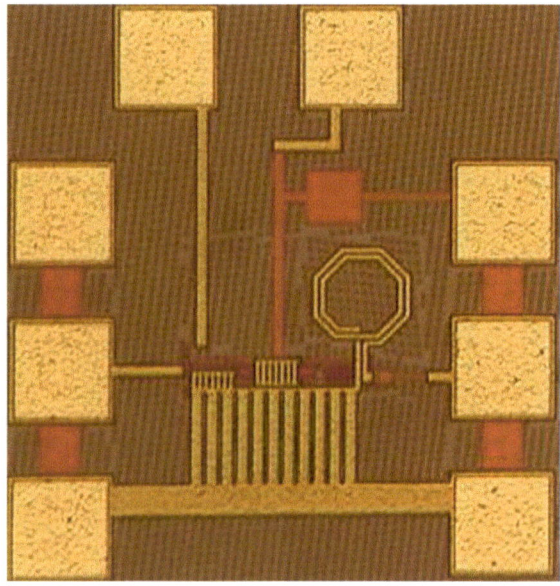

Figure 3.10 shows a photograph of the tunable CMOS monocycle pulse generator. The core of monocycle pulse generator only occupies an area of 240 µm × 160 µm. The CMOS tunable monocycle pulse generator circuit and other accessory components were measured on-wafer in both time and frequency domains using a probe station, digitizing oscilloscope, and spectrum analyzer.

To verify the design of the tunable square-wave generator and pulse shaping circuit, which are parts of the tunable monocycle pulse generator, these constituent components were also fabricated and measured. All the measurements were performed under the condition of 50-Ω load unless otherwise specified.

First, the square-wave generator integrated with the tunable delay cell was measured with a 10-MHz sinusoidal clock signal as the input. The capacitor value of the shunt-capacitor M2 of the tuning delay cell was optimized through simulation with ADS based on the inverters' parameters shown in Table 3.1 to achieve 500 ps around delay tuning range. The final capacitor value of the shunt-capacitor M2 was chosen as 0.2 pF in the simulation. The transistor parameters of the inverters in the corresponding square-wave generator are shown in Table 3.1.

Table 3.1 Transistor
parameters of the inverters in
the square-wave generator

	First-stage inverter		Last-stage inverter	
Transistor	NMOS	PMOS	NMOS	PMOS
Width (µm)	5	15	80	240
Length (µm)	0.18	0.18	0.18	0.18

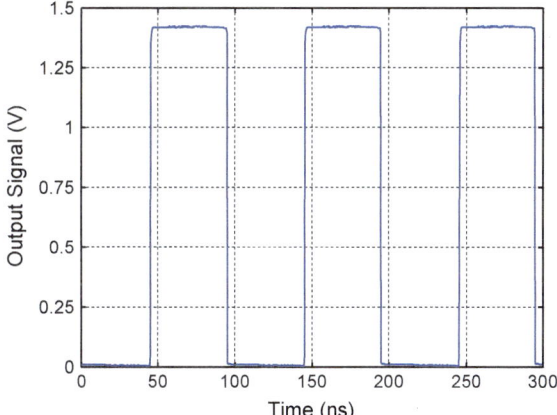

Fig. 3.11 Measured output signal of the square-wave generator

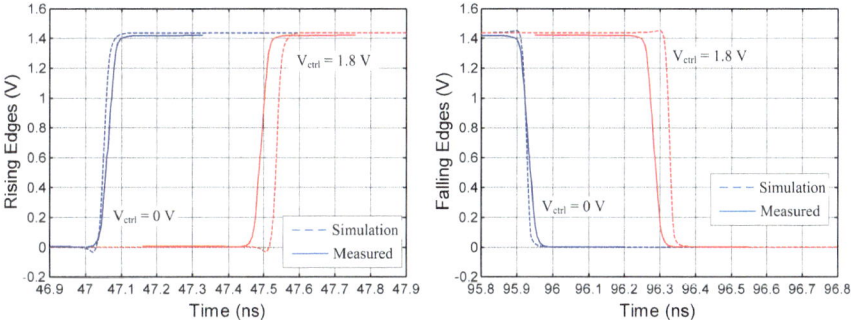

Fig. 3.12 Measured and simulated rising and falling edges of the tunable square-wave signal

The measured output signal of the tunable square-wave generator is shown in Fig. 3.11 with the period of 100 ns. The measured and simulated rising and falling edges of the generated square-wave signal of the tunable square-wave generator corresponding to the delay tuning voltage V_{ctrl} of 0 and 1.8 V, respectively, are shown in Fig. 3.12.

As shown in Fig. 3.11, the waveform generated by the tunable square-wave generator is symmetric and has the good square-wave shape, which validates the design of this component. In addition, Fig. 3.12 presents important information about the delay tuning range and rising/falling edges, which are the critical factors in the tunable pulse generator design. As shown in Fig. 3.12, the measured tuning range of this tunable square-wave generator is around 400 ps, which is slightly narrower than the simulated one. Since the delay tuning range is proportional to the ratio of the shunt-capacitor and the input-capacitor values of the first-stage inverter of the square-wave generator, the parasitic capacitor associated with the inverter makes the overall input-capacitor value larger than the simulated one, which results in reduced capacitor ratio and hence, smaller tuning range. However, the 400-ps delay tuning range still can meet the requirement of the tunable pulse generator design. The details of the rising and falling edges in Fig. 3.12 also confirm the symmetry of the generated square wave. Comparing with the simulation results, the measured rising and falling edges (10–90%) have a width of around 40 ps, which is close to the simulated results. The difference between the two results is caused by the parasitic capacitors associated with the stage inverters of the square-wave generator. The parasitic resistance makes the high-level voltages of the measured results slightly lower than the simulation results. Because of very compact structure, extremely short interconnecting lines, and large enough vias used in the circuit, the resulting parasitic resistance is not large, thereby the difference is not much.

Next, the performance of the pulse shaping circuit was measured in both frequency- and time-domain. The designed shunt spiral inductor has an inductance of 0.53 nH and the series MIM capacitor has 0.4-pF capacitance. Figure 3.13 shows the simulated transfer characteristics of the pulse shaping circuit acting as a high pass filter (HPF). As seen in Fig. 3.13, the pulse-shaping circuit effectively attenuates the low-frequency components of the input signal below 3 GHz.

The measured time-domain results with input impulse signal of different pulse widths are shown in Fig. 3.14, which confirms the design. A commercial pulse generator was used to generate the positive 1-V_{p-p} impulse signal with different pulse widths of 100, 200, and 300 ps. The measured output signals from the pulse-shaping circuit are clearly monocycle pulse signals with amplitude of 0.8 V_{p-p} and almost symmetric positive and negative shapes. Therefore, the designed pulse-shaping circuit can work effectively in the tunable pulse generator to generate the desired tunable monocycle signal when an impulse signal having pulse width varied from 100 to 300 ps is used as the input signal.

To verify the design concept for generating tunable impulses, a separate chip without the pulse-shaping circuitry was first fabricated and measured. The transistor

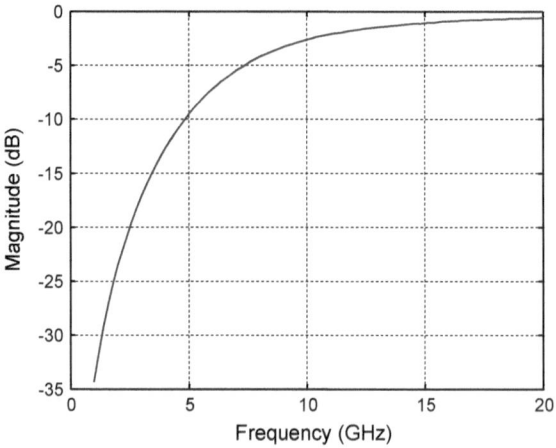

Fig. 3.13 Simulated transfer function of the designed pulse shaping circuit

Fig. 3.14 Measured performance of the pulse-shaping circuit with impulse signal input

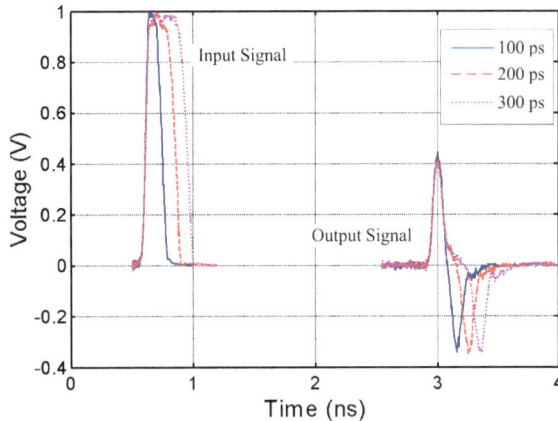

Table 3.2 Transistor parameters of the NOR gate block

Transistor	NMOS	PMOS
Width (μm)	80	320
Length (μm)	0.18	0.18

parameters of the impulse-forming component of the circuit, i.e. NOR gate block as shown in Fig. 3.3, are given in Table 3.2. Large-size transistors were selected to provide enough driving capability for an external 50-Ω load. To reduce the effects of the parasitic capacitor and resistor, multiple-finger gate structure was used for all the transistors of the circuit to improve the high-frequency performance and output power of the generated impulse signal.

The measured and calculated impulse signals with different durations are shown in Fig. 3.15 for a 50-Ω load condition, and are expectedly similar to the voltage waveforms at node C illustrated in Fig. 3.9. Impulse signals having 0.95–1.05 V peak-to-peak voltage with 100–300 ps tunable pulse duration were measured. The pulse duration is defined at 50% of the peak amplitude. The pulse-width tunability is achieved by varying the gate control voltage V_{ctrl} of the tunable delay cell within the range of 0 V to V_{dd}. Figure 3.15 also shows clearly that the generated impulse signals have a common rising edge, whose position is determined only by the falling edge of the square wave at node B in Fig. 3.9; while the position of the falling edge of the generated impulses is determined by the rising edge of the square wave at node A of Fig. 3.9 and the tunable relative time offset between nodes A and B. It is noted that the measured waveforms are very symmetrical with almost no distortion. Good symmetry and low distortion are important for most applications involving pulse signals. Also, as can be seen, the measured results are well matched to the simulated ones. Comparing to the previous work using TSMC 0.25-μm in [9], the tuning duration range with constant pulse amplitude improves significantly.

It should note that the final impulse signal generated generally consists of three parts: rising edge, tunable relative time offset, and falling edge, as shown at node C of Fig. 3.9. For impulses with very narrow pulse widths, only parts of the rising

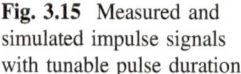

Fig. 3.15 Measured and
simulated impulse signals
with tunable pulse duration

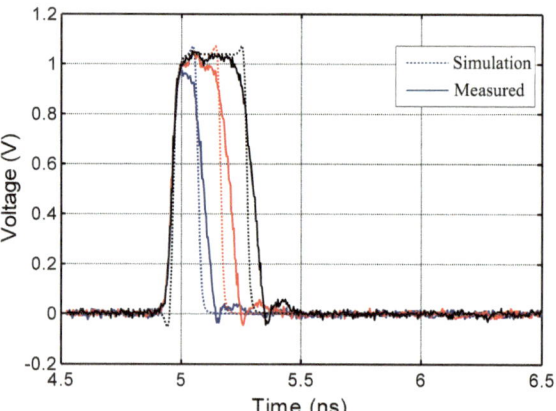

and falling edges of the square waves are involved in the pulse formation, resulting in amplitudes much smaller than those for wider pulses. When the pulse width reaches a certain value, the full rising and falling edges of the square waves and tunable relative time-offset part all contribute to the pulse generation, so the amplitude of the generated impulse signal does not change anymore, and different tuning relative time-offsets will only change the final pulse width. Consequently, there is not much difference in amplitude for different impulse signals if the pulse width exceeds a certain value. For the situation where a impulse signal with minimum pulse width is desired, there is a compromise between the minimum pulse width and signal amplitude, since a very narrow pulse would sacrifice too much pulse energy. Using a better technology such as 0.13-μm CMOS process would improve the tuning range of pulses with uniform amplitude and the amplitude of the minimum-width pulse, because a better process would achieve much sharper rising/falling edges of the square wave, which means a pulse with much narrower minimum width. Hence the corresponding pulse tuning range with constant pulse amplitude is extended.

To verify the frequency response performance of the generated impulse signals, the power spectral density (PSD) of the impulse signal was also measured using a spectrum analyzer, which can cover the frequency range of 9 kHz to 22 GHz. Figure 3.16 displays the measured PSD of the impulse signals with 100-ps and 300-ps pulse durations. The measured results clearly show that, for impulse signals, major PSD components always concentrate within a low-frequency range from dc, which define the bandwidth or frequency operating range of an impulse signal. Accordingly, the bandwidth of the generated impulse signal can be changed by simply tuning the control voltage of the delay cell. For the 100-ps impulse shown in Fig. 3.16a, the first null frequency of the PSD appears at 8 GHz, while that for the 300-ps impulse in Fig. 3.16b occurs around 3.5 GHz. These results verify the design of the tunable impulse generator. The tunable impulse generator can be used further to generate tunable monocycle pulses.

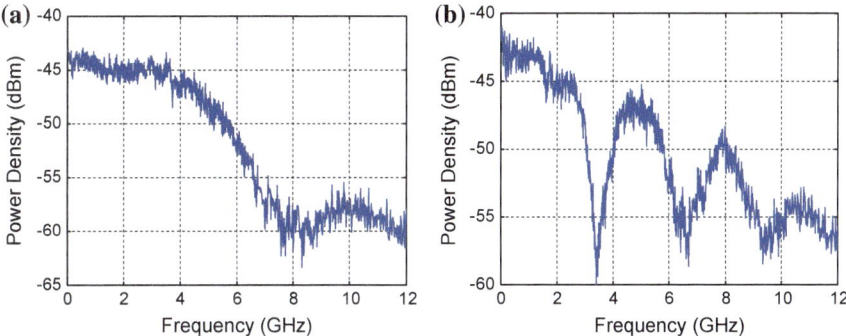

Fig. 3.16 Measured PSD of the tunable impulse signal with pulse width of **a** 100 ps and **b** 300 ps

The relation of the impulse's width to the tuning delay control voltage V_{ctrl} of the tunable impulse generator was also investigated, and the result is presented in Fig. 3.17. For tuning voltages below 0.6 V, the transistor M1 was actually "off", hence functioning as a very large resistor. The equivalent capacitor of the tunable delay cell therefore could be ignored, and the very short relative time-offset between the two paths produced an impulse signal with very low amplitude that is not useful for UWB applications. When the tuning voltage was increased from 0.7 to 1.1 V, the impulse's width increased linearly from 45 to 90 ps, and the corresponding amplitude of the impulse signal increased from 0.4 to 0.9 V. Further increase of the tuning voltage to 1.6 V only widened the impulse's width from 90 to 340 ps with much faster width variation, but the increase of the amplitude of the impulse signal is very little, and hence the amplitude can be considered as constant. That's because the transistor M1 entered the saturation region, and the value of the corresponding equivalent shunt capacitor does not change anymore. As shown in Fig. 3.17, the impulse width variation from 1.6 to 1.8 V is not much. For design that requires the

Fig. 3.17 Measured pulse width of the impulse signal versus tuning delay voltage

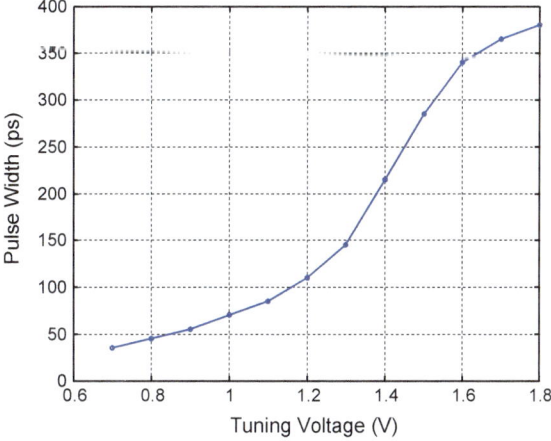

Table 3.3 Transistor
parameters of the NAND gate
block

Transistor	NMOS	PMOS
Width (μm)	80	240
Length (μm)	0.18	0.18

pulse width variation proportional with the tuning voltage change, the shunt NMOS capacitor can be replaced with other components having an extended linear tuning range to meet the requirement.

As mentioned earlier, when the NAND gate block is used as an impulse-forming component in a tunable impulse generator, a tunable impulse signal with negative amplitude can be generated. To demonstrate this, the corresponding tunable negative impulse generator was also fabricated separately and tested following the same previously mentioned test conditions. Table 3.3 presents the parameters of the transistors used in the corresponding NAND gate block. Figure 3.18 shows the measured results of the generated negative tunable pulse signals. As shown in Fig. 3.18, three impulse signals with pulse widths of 100, 200, and 300 ps share the common falling edge with the amplitude ranged from 1 to 1.2 V. The negative impulse signals also maintain good symmetric shapes.

Finally, the measured tunable monocycle pulse signals are shown in Fig. 3.19 for 50 Ω-load condition. By changing the gate control voltage V_{ctrl} of the tunable delay cell in the range of 0 V to V_{dd}, symmetric monocycle pulses with 0.7–0.75 V peak-to-peak voltage and 140–350 ps tunable pulse duration, at 50% of the peak amplitude, were measured, which are also similar to the pulse shapes at node D of Fig. 3.8 as expected. To verify the frequency-response performance of the generated monocycle pulses, their PSD was also measured using a spectrum analyzer. Figure 3.20 displays the measured PSD of the monocycle pulse with 140-ps pulse duration, showing that most of the PSD is below −50 dBm over the 3.1–10.6 GHz UWB frequency band.

Fig. 3.18 Measured negative
impulse signals with tunable
pulse duration (NAND gate
block)

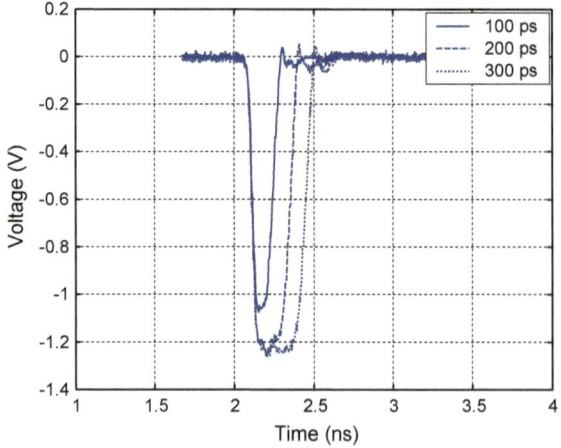

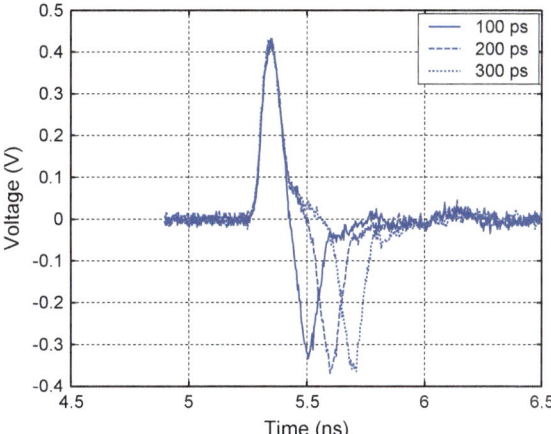

Fig. 3.19 Measured tunable monocycle pulses

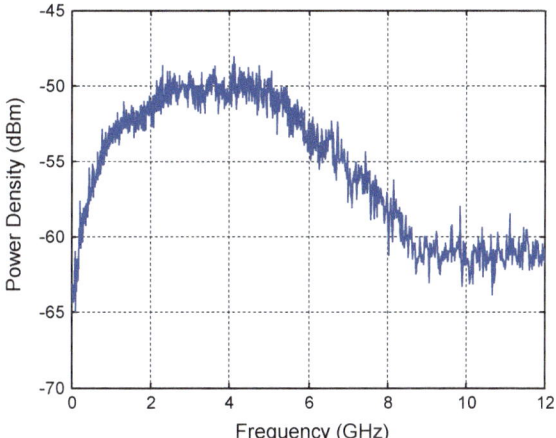

Fig. 3.20 Spectrum of the 140-ps monocycle pulse signal

3.4 BPSK Modulator Design

For impulse-type UWB transmitter design, BPSK modulation is normally chosen to modulate a digital information data sequence to a pulse sequence [7, 19]. For the BPSK modulation, the polarities of the output pulse signals can be controlled by the polarities of the information data levels. The BPSK has an advantage over the pulse amplitude and position modulation due to two times improvement in the overall power efficiency [20], an inherent 3-dB increase in separation between the constellation points. In this section, a simple level triggered pulse modulation circuit is developed to achieve the BPSK modulation, which is fully integrated with the tunable pulse generators described in the previous sections.

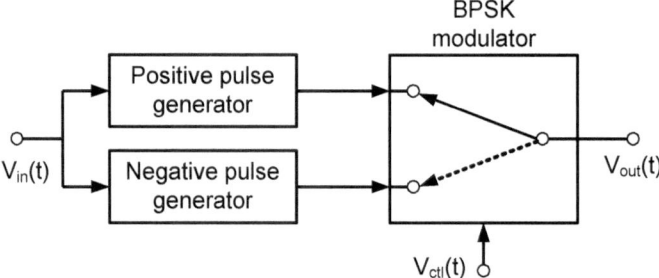

Fig. 3.21 BPSK block diagram

The block diagram of the BPSK pulse modulator is shown in Fig. 3.21, which includes two pulse generators and one switch. A 10-MHz clock signal is used as the input signal $V_{in}(t)$ of the pulse generators. This pulse generator can produce both positive and negative tunable pulse signals. The control input signal $V_{ctl}(t)$ to the BPSK modulator is the (transmitting) information data sequence to be transmitted, and the level of this modulation signal determines the polarities of the final output pulse signal $V_{out}(t)$. When the digital modulation signal is at low level, such as "0", the output signal of the BPSK modulator will be a pulse signal with negative amplitude; while for the digital information signal with high level of "1", a positive pulse signal will be generated at the output of the BPSK modulator.

Figure 3.22 shows the BPSK modulation circuit to be implemented in CMOS technology, which consists of R_1, R_2, C_1, C_2, M_1, M_2, and M_3. V_{inp} is the input pulse signal with positive amplitude, and V_{inn} is the input pulse signal with negative amplitude. V_{ctl} is the input digital information data sequence to be transmitted, which can be a low level "0" or a high level "1". V_{out} is the output signal of the BPSK modulator, which is loaded by an external 50-Ω resistor (not shown in the

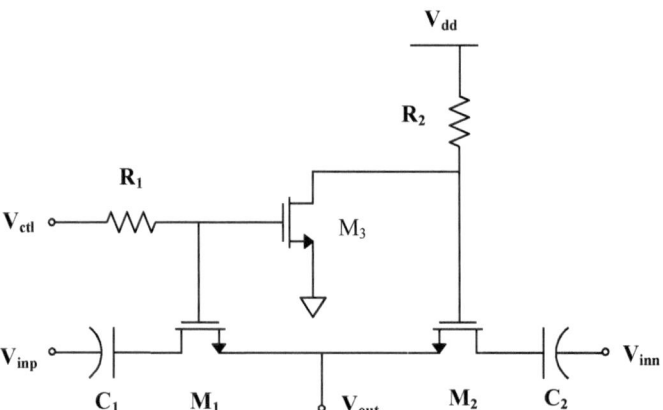

Fig. 3.22 BPSK modulation circuit

Table 3.4 Parameters of the transistors in the BPSK modulator

Transistor	M1	M2	M3
Width (μm)	160	160	20
Length (μm)	0.18	0.18	0.18

circuit). NMOS transistors M_1, M_2, and M_3 together form a multiplexer, where M_1 and M_2 are used as two transmission gates and biased by complimentary control voltages controlled by M_3. Thus, at one time, only one input signal can pass through the transmission gate M_1 or M_2, and feed to the 50-Ω load to generate a modulated positive or negative pulse.

As shown in Fig. 3.22, when the modulating digital data is at high level "1", the corresponding control signal V_{ctl} is V_{dd}, hence the gate voltage of both M_1 and M_3 is V_{dd}. Therefore, the transmission gate M_1 is in the "on" condition, and transistor M_3 is "on" in the saturation region. Hence the gate voltage of M_2 is close to zero. This makes the transmission gate M_2 in the "off" condition. So only the positive pulse signal V_{inp} is passed to the output of BPSK modulator. For the digital data information of low level "0", the corresponding control signal V_{ctl} is zero, hence the gate voltage of both M_1 and M_3 is zero. So the transmission gate M_1 and M_3 are both "off", and the gate voltage of M_2 is close to V_{dd}, resulting the "on" condition for M_2, and the negative pulse signal V_{inn} is generated at the output of the BPSK modulator.

For the BPSK modulator, the transmission gates M_1 and M_2 should drive the 50-Ω load effectively. As larger transmission-gate transistors mean smaller on-resistance [14], the transmission gates M_1 and M_2 should be selected sufficiently large to avoid amplitude degradation. Table 3.4 provides the parameters of the corresponding NMOS transistors.

Figures 3.23 and 3.24 present the simulated performance of the BPSK modulator in the frequency- and time-domain. Less than 1.5 dB insertion loss is achieved over the entire UWB band, while the isolation is less than 20 dB over most of that frequency range. As shown in Fig. 3.24, the output signals retain the same shape as the input pulses for both the "1" and "0" conditions.

Fig. 3.23 Simulated insertion loss and isolation of then BPSK modulator

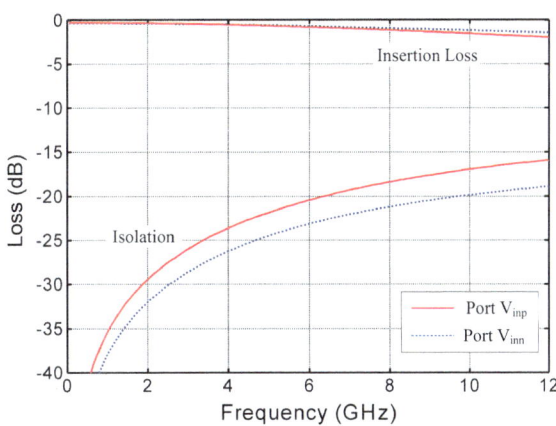

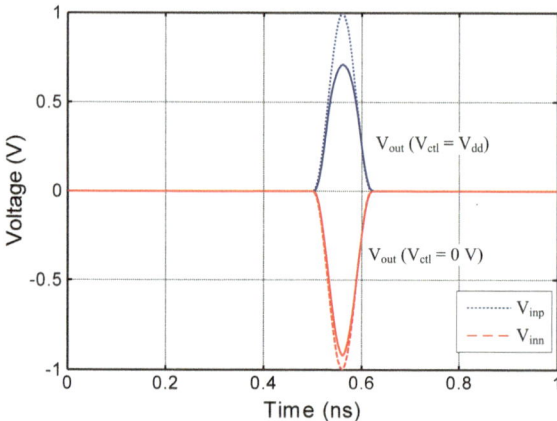

Fig. 3.24 Simulated time-domain performance of the BPSK modulator

3.5   UWB Tunable Transmitter Design

In this section, the tunable impulse and monocycle pulse UWB transmitters, integrating the corresponding tunable impulse and monocycle pulse generators with the BPSK modulator in a single CMOS chip, are presented.

3.5.1   UWB Tunable Impulse Transmitter Design

The UWB tunable impulse transmitter's block diagram is shown in Fig. 3.25, which integrates the tunable impulse generator, CMOS inverter, and BPSK modulator together. As described earlier, the tunable impulse generator driven by the 10-MHz input clock signal produces the impulse with tunable duration, which is divided equally into two paths: one impulse signal is directly sent to the BPSK modulator, functioning as the positive impulse input signal V_{inp} as shown in Fig. 3.22; another signal goes through the CMOS inverter and then to the BPSK modulator to generate the negative pulse signal, hence feeding the impulse signal V_{inn} with negative amplitude to the modulator as seen in Fig. 3.22. Considering the extremely narrow rising and falling edges of the impulse signal, a large size for the CMOS inverter was chosen to provide enough driving capability for the impulse,

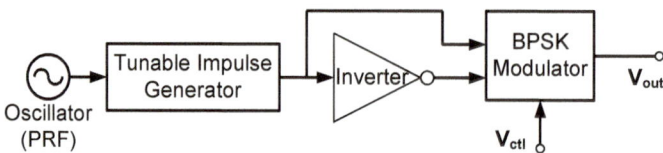

Fig. 3.25 Block diagram of the UWB tunable impulse transmitter

Fig. 3.26 Photograph of the
UWB tunable impulse
transmitter

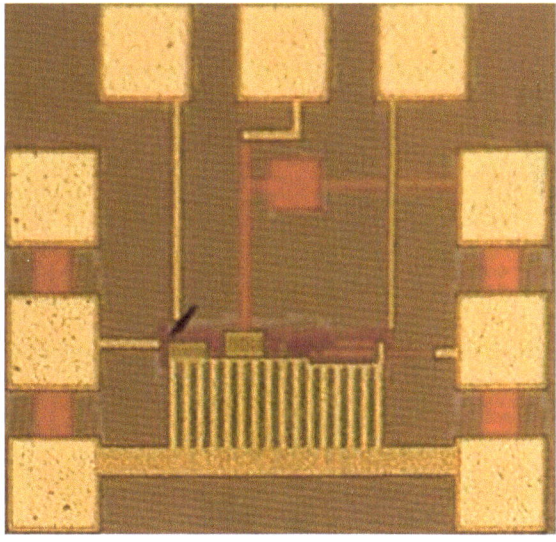

otherwise the generated negative impulse V_{inn} at the output of the CMOS inverter
would degrade the pulse width and amplitude performance. The output signal V_{out}
of the transmitter is determined by the external modulating signal V_{ctl} to generate
the tunable impulse signal with positive or negative amplitude.

Figure 3.26 shows a photograph of the UWB tunable impulse transmitter fabricated with Jazz 0.18-μm RFCMOS technology [16]. The overall size of the circuit
is 580 μm × 550 μm including the RF pads and dc pads for on-wafer measurement
purpose. The external modulating signal was fed to the on-wafer dc pad of the
circuit through dc probe. The measurement results of the output modulated impulse
signals are shown in Fig. 3.27 for different modulation signals, where $V_{ctl} = V_{dd}$ is
the high level, and $V_{ctl} = 0$ is the low level.

Figure 3.27a shows the positive tunable impulse signals with pulse widths of
100 and 300 ps for the high-level modulation signal. The measured pulse widths
and shapes are comparable with the those simulated, while the measured amplitudes
are somewhat lower than the simulated ones, which were caused by extra parasitic
resistances from interconnection lines that could not be fully included in the simulations. Comparing with the impulse signals shown in Fig. 3.15 generated by the
tunable impulse generator, the modulated impulse signals have a smaller amplitude
of 0.8 V, yet still keeping symmetrical shapes and similar pulse widths. For the
negative impulse signals shown in Fig. 3.27b under the condition of low-level
modulation, the amplitudes are reduced to around 0.7 V. Compared to the measured
positive pulses, the measured negative pulses experience more distortions over a
part of the pulse durations at the (bottom) negative parts of the pulse waveforms.
The reason is that the input negative impulse to the modulator is produced by the
CMOS inverter, which attenuates the pulse amplitude and expands the pulse width
to some extent. A possible way to lessen this effect is to replace the CMOS inverter

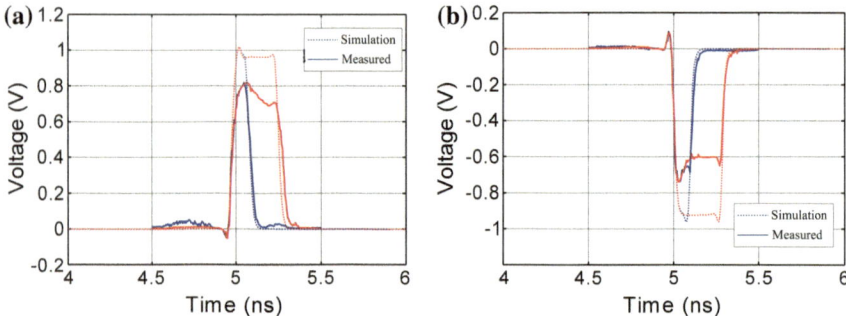

Fig. 3.27 Measured and simulated output impulse signals of the CMOS UWB tunable impulse transmitter for **a** high- and **b** low-level modulating signals

that generates the negative impulse with the NAND gate block based negative impulse generator, which would provide a good negative impulse signal as shown in Fig. 3.18 with large enough amplitude to the BPSK modulator. The consumed power, however, will also increase accordingly.

For the UWB tunable monocycle-pulse transmitter that integrates the monocycle pulse generator with the BPSK modulator, two circuit topologies can be used. One is having the BPSK modulator succeeding two monocycle pulse generators that can generate two identical monocycle pulse signals with opposite polarities, which is a straightforward way to produce the modulated monocycle pulse signals. However, using two monocycle pulse generators would occupy much die space and suffer increased power consumption caused by the two monocycle pulse generators. The other configuration is realized by placing a pulse-shaping module right after the UWB impulse transmitter described earlier. With this scheme, the modulated impulse signal with positive or negative amplitude is shaped into the modulated monocycle pulse signal with corresponding positive or negative amplitude. Since only a pulse-shaping component is used instead of two monocycle pulse generators with the other topology, the die area of the UWB tunable monocycle-pulse transmitter is substantially smaller. Hence, this topology is selected and the final structure of the UWB monocycle-pulse transmitter is shown in Fig. 3.28. The pulse-shaping component as mentioned previously is essentially a high-pass filter It is noted that the CMOS monocycle pulse generator with its performance described earlier in Figs. 3.6, 3.10 and 3.19 is not used in the CMOS UWB monocycle-pulse transmitter, and their design and performance are indeed different.

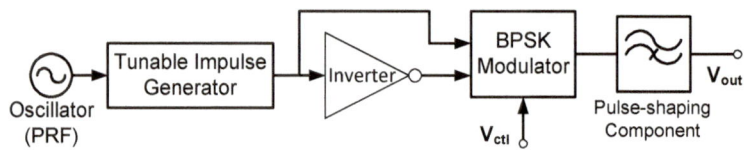

Fig. 3.28 Block diagram of the UWB tunable monocycle-pulse transmitter

The UWB monocycle-pulse transmitter, as shown in Fig. 3.28, consists of the tunable impulse generator, CMOS inverter, BPSK modulator, and pulse-shaping component. The 10-MHz clock signal drives the tunable impulse generator to produce the positive and negative impulses with tunable durations. The positive impulse signal generated directly from the tunable impulse generator and the negative impulse signal emerged from the CMOS inverter are fed to the BPSK modulator. Depending on the external high- or low-level modulating signal V_{ctl}, the positive (high-level modulation) or negative (low-level modulation) modulated impulse output of the BPSK modulator is sent to the pulse-shaping circuit. Finally, the modulated positive or negative monocycle pulse signal is produced at the output of the pulse-shaping circuit.

A photograph of the UWB monocycle-pulse transmitter with BPSK modulator is shown in Fig. 3.29, which was fabricated with Jazz 0.18 μm RFCMOS technology [16]. The overall size of the circuit is 620 μm × 550 μm including the on-wafer RF and dc pads. The measured and simulated output modulated monocycle pulse signals are shown in Fig. 3.30 for different modulation signals where $V_{ctl} = V_{dd}$ is the high level and $V_{ctl} = 0$ is the low level.

As shown in Fig. 3.30, the measured modulated monocycle pulse signals with pulse width of 100 and 300 ps have the peak-to-peak amplitude of 0.6–0.8 V, and the symmetry of the signal shape is somewhat degraded as compared with the simulation results, while the pulse widths remain the same. The pulse shape's unsymmetrical problem is mainly caused by the pulse-shaping component succeeding the BPSK modulator.

Fig. 3.29 Photograph of the UWB monocycle-pulse transmitter

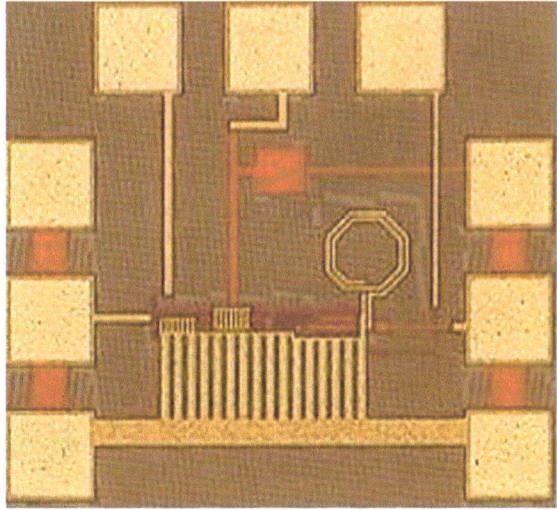

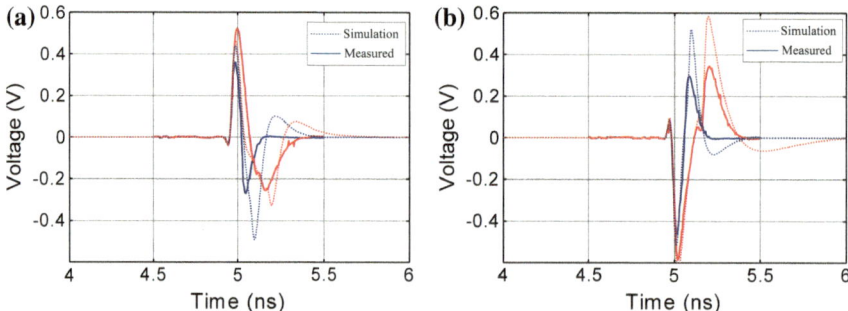

Fig. 3.30 Measured and simulated output pulses of the UWB monocycle-pulse transmitter for **a** high- and **b** low-level modulating signals

3.6   Summary

This chapter presents the design of CMOS UWB tunable sub-nanosecond impulse and monocycle-pulse BPSK transmitters. The UWB impulse transmitter produces positive and negative tunable impulse signals corresponding to the high- and low-level modulation signals, respectively. The positive and negative impulse signals have amplitudes of 0.8 and 0.6 V with tunable pulse widths from 100 to 300 ps, respectively. The UWB monocycle-pulse transmitter produces monocycle pulse signals of opposite polarities having peak-to-peak amplitudes of about 0.6– 0.8 V and tunable pulse widths between 100 and 300 ps. This chapter also covers the design of the CMOS impulse generator and BPSK modulator used in the UWB impulse and monocycle-pulse transmitters. The impulse generator can generate 0.95–1.05 V peak-to-peak Gaussian-type impulse signal with 100–300 ps tunable pulse duration. Moreover, this chapter includes the design of a CMOS monocycle pulse generator, which can produce 0.7–0.75 V peak-to-peak monocycle pulse with 140–350-ps tunable pulse duration.

References

1. J.W. Han, C. Nguyen, Development of a tunable multi-band UWB radar sensor and its applications to subsurface sensing. IEEE Sens. J. **7**(1), 51–58 (2007)
2. J.W. Han, C. Nguyen, On the development of a compact sub-nanosecond tunable monocycle pulse transmitter for UWB applications. IEEE Trans. Microw. Theory Tech. **MTT-54**(1), 285–293 (2006)
3. J. Han, C. Nguyen, Ultra-wideband electronically tunable pulse generators. IEEE Microw. Wirel. Compon. Lett. **14**(3), 112–114 (2004)
4. S. Bagga, W.A. Serdijn, J.R. Long, A PPM Gaussian monocycle transmitter for ultra-wideband communications, in *Proceedings of Joint UWBST & IWUWBS*, pp. 130–134, May 2004

5. Y. Jeong, S. Jung, J. Liu, A CMOS impulse generator for UWB wireless communication systems, in *IEEE International Symposium on Circuits and Systems (ISCAS 2004)*, vol. 4, pp. 129–132 (2004)
6. H. Kim, Y. Joo, Fifth-derivative Gaussian pulse generator for UWB system, in *IEEE Radio Frequency Integrated Circuits (RFIC) Symposium*, pp. 671–674, June 2005
7. Y. Zheng, H. Dong, Y.P. Xu, A novel CMOS/BiCMOS UWB pulse generator and modulator, in *IEEE International Microwave Symposium Digest*, pp. 1269–1272, June 2004
8. I. Oppermann, M. Hamalainen, J. Iinatti, *UWB Theory and Applications* (Wiley, Hoboken, NJ, 2004)
9. M. Miao, C. Nguyen, On the development of an integrated CMOS-based UWB tunable–pulse transmit module. IEEE Trans. Microw. Theory Tech. **MTT-54**(10), 3681–3687
10. M. Miao, C. Nguyen, Fully integrated CMOS impulse UWB transmitter front-ends with BPSK modulation. Microwave Opt. Technol. Lett. **52**(7), 1609–1614 (2010)
11. S.M. Kang, Y. Leblebici, *CMOS Digital Integrated Circuits: Analysis and Design*, 3rd edn. (McGraw-Hill Higher Education, New York, NY, 2003)
12. M.G. Johnson, E.L. Hudson, A variable delay line PLL for CPU-coprocessor synchronization. IEEE J. Solid-State Circuits **23**(5), 1218–1223 (1988)
13. M. Maymandi-Nejad, M. Sachdev, A digitally programmable delay element: design and analysis. IEEE Trans. VLSI Syst. **11**(5), 871–878 (2003)
14. R.J. Baker, *CMOS Circuit Design, Layout, and Simulation*, 2nd edn. (IEEE Press, Piscataway, NJ, 2005)
15. IE3D. Zeland Software Inc., Fremont, CA (2006)
16. Jazz 0.18-μm CMOS Process. Jazz Semiconductor, Newport Beach, CA (2006)
17. Advanced Design System. Agilent Technologies Inc., Santa Clara, CA (2006)
18. Cadence Design Systems. Cadence, San Jose, CA (2006)
19. D.D. Wentzloff, A.P. Chandrakasan, Gaussian pulse generators for subbanded ultra-wideband transmitters. IEEE Trans. Microw. Theory Tech. **54**, 1647–1655 (2006)
20. H. Arslan, Z.N. Chen, M.G. Di Benedetto, *Ultra Wideband Wireless Communications* (Wiley, Hoboken, NJ, 2006)

Chapter 4
UWB Receiver Design

4.1 Introduction

A UWB impulse receiver receives UWB pulse signals through its receiving antenna and down-converts them into a baseband signal. For efficient conversion, the down-converted baseband signal should preserve the waveform as close to that of the received signal as possible. Since the received pulse signal covers an ultra-wide band such as 3.1–10.6 GHz, the design of an impulse receiver that can down-convert an input pulse signal and maintain faithfully its waveform is challenging.

As discussed in Chap. 2, the architecture of the UWB impulse receiver is much simpler than those of conventional narrow-band systems and MB-OFDM UWB receivers. It only consists of a UWB LNA, down-conversion mixer (or correlator), and template pulse generator. Among these components, the UWB LNA and correlator are the two essential circuits. The performances of the LNA and correlator directly determine the final performance of the UWB impulse receiver. In UWB impulse systems, no matter what kind of modulation technique is used, the correlator and LNA with minimum NF are always indispensable components for the receiver. To reduce the cost, size and power consumption of UWB wireless systems, especially mobile devices, it is necessary to integrate all the UWB components in a single CMOS chip with good performance for the UWB impulse receiver. Considering the ultra-wide bandwidth of UWB from 3.1 to 10.6 GHz, this requirement represents a challenge for circuit design.

In this chapter, a UWB impulse receiver implemented on CMOS technology [1] is presented with particular focus on its core components of UWB LNA and correlator. Specifically, individual UWB LNA and correlator circuits are designed, fabricated, and measured to verify the design topology. The UWB impulse receiver front-end integrating the UWB LNA, correlator, and template pulse generator is then presented.

© The Author(s) 2017   57
C. Nguyen and M. Miao, *Design of CMOS RFIC Ultra-Wideband Impulse Transmitters and Receivers*, SpringerBriefs in Electrical and Computer Engineering, DOI 10.1007/978-3-319-53107-6_4

4.2 UWB LNA

4.2.1 UWB LNA Topology

A wideband LNA operating over the whole UWB band of 3.1–10.6 GHz is an important component in both UWB impulse and MB-OFDM receivers. This amplifier should have wideband matching, high gain, good linearity, low noise figure (NF) over the entire bandwidth, and low power consumption. A LNA was designed and implemented over the entire UWB band with Jazz 0.18 μm RFCMOS process. To facilitate the measurement, an output buffer was also included with the LNA to drive an external 50 Ω load. The LNA without the buffer is integrated with the UWB correlator to form the impulse-type UWB receiver.

There are many different circuit configurations for high-frequency wideband amplifiers depending on requirements and applications. Figure 4.1 shows typical structures including shunt feedback, distributed, common-gate (CG), and cascoded common-source (CS) topologies [2]. To facilitate the selection of a proper UWB LNA topology for our intended UWB application, we will first briefly describe the advantages and disadvantages of these topologies regarding power consumption, NF, and die area and a proper topology is then chosen for the desired UWB LNA.

4.2.1.1 Shunt Feedback LNA

In shunt feedback amplifiers, the negative feedback achieves simultaneous impedance match at both input and output ports, and produces relatively constant input and output impedances over a broad frequency range [2]. However, as can be inferred in Fig. 4.1a, the inherently large gate-source capacitance (C_{gs}) of the CMOS transistor results in a large parasitic input capacitance, causing limited input impedance match bandwidth at high frequencies [3]. Furthermore, the resistive feedback network generates its own thermal noise, which results in the amplifier's overall NF to generally exceed the device's minimum NF (NF_{min}) by a considerable

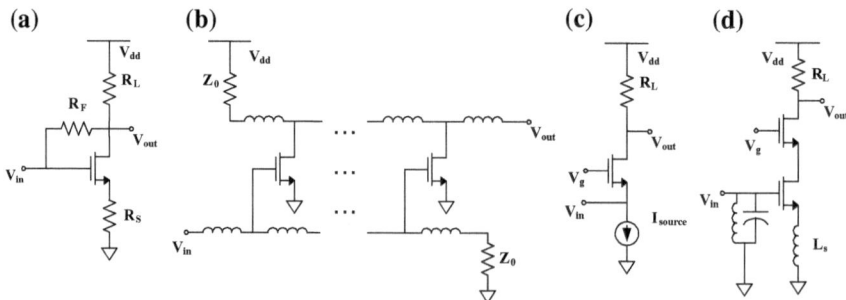

Fig. 4.1 Typical wideband LNA topologies: **a** shunt feedback, **b** distributed amplifier, **c** common-gate, and **d** cascoded common-source

amount [2]. Therefore, the shunt feedback structure may not provide sufficiently low NF and high gain while consuming low power, which is an important specification in the UWB LNA design.

4.2.1.2 Distributed LNA

In contrast with typical multi-stage amplifiers consisting of cascaded stages, the overall gain of the distributed amplifiers, which also encompass multiple stages, depends linearly on the number of the gain stages (to a certain limit). Hence, the distributed amplifiers in general may operate at substantially higher frequencies with ultra wideband performance than the shunt feedback amplifiers. However, the cost of the power consumption in the distributed structures is relatively large as compared to the feedback structure due to multiple gain stages. Moreover, as shown in Fig. 4.1b, the number of inductors used in the distributed structure results in a large die area and hence high cost, making this type of amplifier less attractive for UWB applications requiring small size and low cost.

4.2.1.3 Common-Gate LNA

The common-gate amplifier seems to be a good choice for UWB systems in terms of power dissipation and die area. As shown in Fig. 4.1c, a wideband input match (to 50 Ω) can be achieved simply by properly selecting the device size and adjusting the bias current such that $1/g_m$, where g_m is the transistor's transconductance, of the amplifying transistor is nearly 50 Ω over a broad frequency range [2]. No input impedance-matching network, therefore, needs to be designed, hence simplifying the overall amplifier design and reducing the die size. However, the lower bound of the NF for common-gate amplifier is about 3 dB [2]. This NF is even worse at high frequencies and when the gate current noise is taken into account. Hence, a presence of noisy resistances in the signal path, such as the inevitable transistor's channel resistance, results in NF degradation [2] and limits the minimum possible NF, which is an adverse effect in UWB systems.

4.2.1.4 Cascoded Common-Source LNA

The cascoded CS topology, as shown in Fig. 4.1d, is often used in wideband LNA design for ease in achieving low NF and high gain. The structure is based on a narrowband inductively degenerated cascoded LNA that is extended to a wide bandwidth by including a band pass filter (BPF) at the input, where the reactive part of the input impedance is resonated over the BPF frequency range [3]. The main feature of this topology is its ability to match the NF close to NF_{min} while also

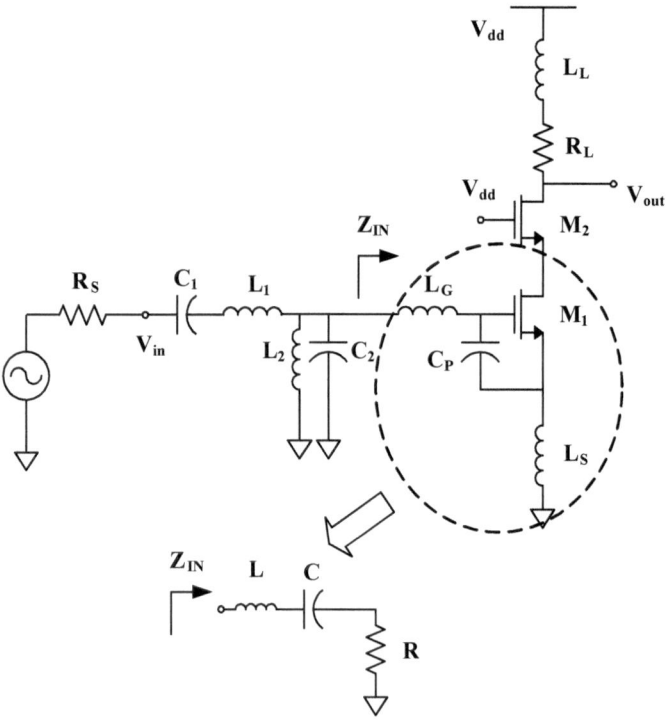

Fig. 4.2 Circuit schematic of the UWB LNA with ladder matching

achieving power match. These characteristics make the cascoded CS topology one of the preferred options for UWB LNA design. This topology, however, has a disadvantage in size, in which the typically used wideband LC matching network contains multiple on-chip spiral inductors, which occupies much die area.

In view of the advantages and disadvantages of the above-described wideband LNA topologies in terms of NF, power dissipation, and die area, the cascoded common-source inductively degenerated LNA with extended ultra-wideband ladder matching network is selected for the UWB LNA of our UWB impulse receiver. This UWB LNA is shown in Fig. 4.2, which is essentially based on the narrow-band cascoded inductively degenerated common-source LNA [2]. The cascoded configuration of the transistors M_1 and M_2 reduces the Miller-effect and improves the input-output reverse isolation as well as frequency response. Because of the reverse isolation achieved by the cascoded structure, the effects of M_2, R_L, and L_L to the input impedance can be neglected.

4.2.2 UWB LNA Analysis

The input impedance of the NMOS transistor M_1 with inductive source degeneration shown in Fig. 4.2 is equivalent to the impedance of a series RLC circuit with R given by [2]

$$R = \omega_T L_S \tag{4.1}$$

where $\omega_T = g_m/(C_{gs} + C_P) = g_m/C_T$ is the cut-off frequency of the transistor. To make the UWB LNA design more flexible, an on-chip spiral inductor L_G is placed in series with the gate of M_1, and an external MIM capacitor C_P is placed in parallel with C_{gs} of M_1.

The input impedance of M_1 with inductive source degeneration can be written as [3]

$$Z_{IN} = \frac{1}{j\omega(C_{gs} + C_P)} + j\omega(L_S + L_G) + \omega_T L_S \tag{4.2}$$

where the real part of Z_{IN} is chosen to be equal to the source resistance R_S, and the reactive part of the input impedance is resonated at the operating frequency with nearly optimal NF [2].

The bandwidth of the narrowband inductively degenerated cascoded LNA is extended by adding the series inductor-capacitor (L_1, C_1) and parallel inductor-capacitor (L_2, C_2) to match the topology of a third-order Chebyshev bandpass filter, as shown in Fig. 4.3, where R is the 50 Ω load. Since the reactive elements of the filter (L_1, C_1, L_2, C_2, L, and C) determine the bandwidth and ripple of the passband, the input reflection coefficient can be written, assuming 0 dB power loss in the passband with ripple ρ, as

$$|\Gamma|^2 = 1 - \frac{1}{\rho} \tag{4.3}$$

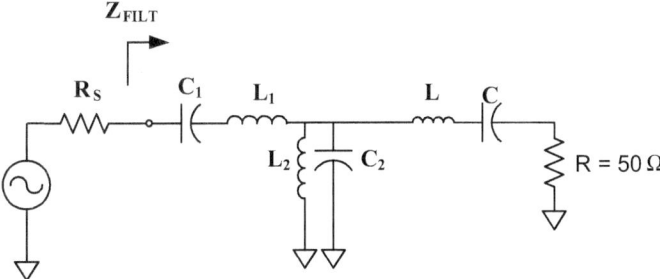

Fig. 4.3 Third-order Chebyshev bandpass filter

Table 4.1 Component values of the third-order Chebyshev BPF

L_1 (nH)	C_1 (pF)	L_2 (nH)	C_2 (pF)	L (nH)	C (pH)
1.65	0.47	1.64	0.47	1.65	0.47

In the passband of the Chebyshev BPF, if the input reflection coefficient is smaller than -10 dB, the tolerable ripple of less than 0.5 dB can be derived from (4.3). For the impulse-type UWB applications, assuming the filter's passband of 3.1–10.6 GHz, the three-section Chebyshev BPF structure is selected considering a compromise between the filter's complexity and component values. The module Filter Design Guide of ADS [4] is employed as the simulation tool to quickly determine the initial ideal component parameters of the three-section Chebyshev bandpass filter for 50 Ω input and output matching, which is shown in Fig. 4.3. The component values of the three-section Chebyshev BPF are presented in Table 4.1, where $L = L_G + L_S$ and $C = C_{gs} + C_P$.

The simulated return loss and insertion loss of the three-section Chebyshev BPF are shown in Fig. 4.4, which cover the UWB band of 3.1–10.6 GHz. The components of three-section Chebyshev BPF will later be replaced with the available process design kit (PDK) on-chip MIM capacitors and electromagnetic (EM) designed and optimized spiral inductors to achieve a fully integrated LNA structure.

As shown in Fig. 4.2, in order to achieve a flat gain over the whole UWB band, the shunt-peaking topology is employed, which includes the series inductor L_L and resistor R_L as the load [2]. The value of L_L should be large enough to provide a large gain at the higher frequency edge, while being so small that the resonating frequency generated by L_L and C_{OUT} is much higher than the operating frequency band, where C_{OUT} is the total capacitance between the drain of M_2 and ground [3]. As for R_L, the zero frequency $\omega_Z = R_L/L_L$ should be close to the lower frequency edge of the band to improve the gain at lower frequencies. R_L is limited by an upper

Fig. 4.4 Simulated performance of the three-section Chebyshev BPF

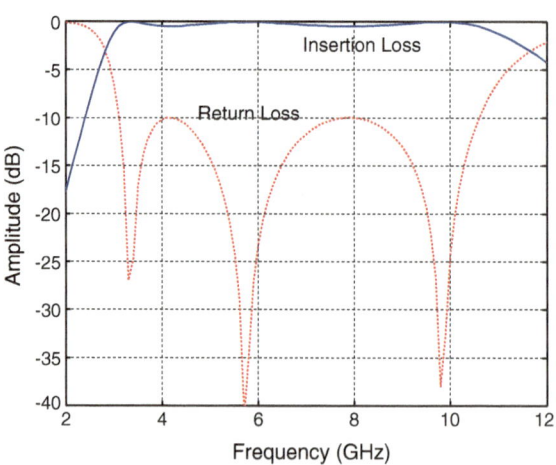

value above which the voltage drop is such large that reduces the voltage V_{dd} supplied to the drain of M2.

We now analyze the voltage gain of the UWB LNA over the entire UWB frequency band. First, for simplicity without loss of generality, we assume the input network of the LNA (i.e., the Chebyshev BPF filter) has a transfer function approximately equal to unity in the passband. The input impedance can be considered over the passband as shown in Fig. 4.2. Hence, the current to the amplifying transistor M_1 can be given as $i_{in} = v_{in}/R_S$. In addition, the CMOS transistor functions as the current amplifier at high frequencies with the current gain of $\beta = g_m/(j\omega C_T)$ [2].

Considering the shunt-peaking load of R_L and L_L, the overall output load can be expressed as

$$Z_{LOAD} = (R_L + j\omega L_L) \left\| \left(\frac{1}{j\omega C_{OUT}} \right) = \frac{R_L + j\omega L_L}{1 + j\omega C_{OUT}(R_L + j\omega L_L)} \right. \tag{4.4}$$

Using (4.4) and the current gain expression $\beta = g_m/(j\omega C_T)$, the overall voltage gain of the amplifier can be written as [3]:

$$\frac{v_{out}}{v_{in}} = -\beta \left(\frac{Z_{LOAD}}{R_S} \right) = -\frac{g_m}{j\omega C_T R_S} \frac{R_L + j\omega L_L}{1 + j\omega C_{OUT}(R_L + j\omega L_L)} \tag{4.5}$$

which clearly shows that, at lower frequencies, R_L plays an important role in the voltage-gain determination, while at higher frequencies, the current gain roll-off is compensated by the load inductor L_L. Furthermore, the spurious resonance introduced by C_{OUT} and L_L has to be kept out of the passband.

4.2.3 UWB LNA Noise Analysis

Two major noise contributors should be considered in the calculation of the UWB LNA's noise performance: the losses associated with the input network and the noise generated by the amplifying transistor M_1. For the input network with MIM capacitors and on-chip spiral inductors as shown in Fig. 4.3, the MIM capacitors normally have much higher quality factors than those of the on-chip spiral inductors. Hence, the noise of the three-section Chebyshev BPF is mainly contributed by the quality factors of the on-chip spiral inductors. To reduce this part of noise, the on-chip spiral inductors, to be described later, are implemented with patterned ground shield and designed and optimized through EM simulations to achieve the highest possible Q for specific inductance values.

For the noise contribution from M_1 for a specific bias current, the transistor's width should be properly selected to achieve the optimum noise value. Considering the ultra-wideband feature in our case, the noise performance of the amplifier over the whole UWB band should be studied. To that end, both the minimum NF and average in-band NF should be investigated and considered for achieving the optimum noise performance for the UWB LNA.

A typical two-port network is represented by the input-referred noise current and voltage source is used in the noise analysis. The $1/f$ noise is ignored because of the amplifier's high operating frequencies. Figure 4.5a shows the MOS transistor's noise sources including the loading effect of the local feedback inductor L_S. $\overline{i_{ng}^2}$ is the induced gate noise due to the coupling of the fluctuating channel charge into the gate terminal, while $\overline{i_{nd}^2}$ is the drain noise current due to the carrier thermal agitation in the channel. The corresponding induced gate noise and drain current noise are expressed respectively as [2]

$$\frac{\overline{i_{ng}^2}}{\Delta f} = 4kT\delta g_g \tag{4.6}$$

$$\frac{\overline{i_{nd}^2}}{\Delta f} = 4kT\gamma g_{d0} \tag{4.7}$$

where k is the Boltzmann's constant, T is the absolute temperature in Kelvin degrees, $g_g = \frac{\omega^2 C_{gs}^2}{5 g_{d0}}$, with g_{d0} being the channel conductance at $V_{DS} = 0$, and $\delta = 1.33 - 4$ and $\gamma = 0.67 - 1.33$ are the excess noise parameters [2].

Employing the conventional input-referred topology in [5], the noise sources of M_1 are replaced with two correlated noise generators of $\overline{i_n^2}$ and $\overline{v_n^2}$, which are shown in Fig. 4.5b [3]. As seen in Fig. 4.5b, when the input is short-ended, only the noise generator $\overline{i_n^2}$ exists. Since the transistor can be considered as current amplifier [2], for the noise source $\overline{i_{nd}^2}$, assuming the input-referred noise current is $\overline{i_{nd,input}^2}$, we have the following relation [2]:

$$\frac{i_{nd}}{\sqrt{\Delta f}} = \frac{g_m}{j\omega C_T} \frac{i_{nd,input}}{\sqrt{\Delta f}} \tag{4.8a}$$

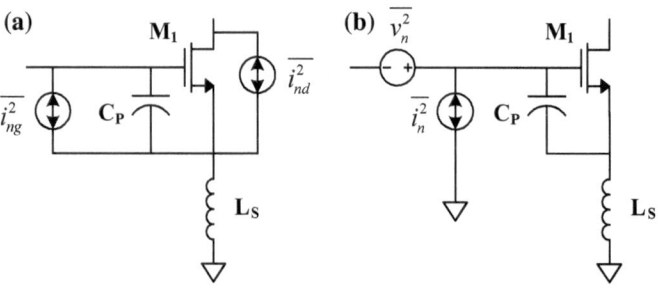

Fig. 4.5 Noise model for transistor M_1: **a** M_1 noise sources and **b** input-referred equivalent noise generators

where $1/j\omega C_T$ is the output impedance caused by the output parasitic capacitor, from which we obtain

$$\frac{i_{nd,input}}{\sqrt{\Delta f}} = \frac{j\omega C_T}{g_m} \frac{i_{nd}}{\sqrt{\Delta f}} \quad (4.8b)$$

Therefore, the total noise current generator $\overline{i_n^2}$ can be derived from (4.8b) as:

$$\frac{i_n}{\sqrt{\Delta f}} = \frac{i_{ng}}{\sqrt{\Delta f}} + \frac{i_{nd,input}}{\sqrt{\Delta f}} = \frac{i_{ng}}{\sqrt{\Delta f}} + \frac{j\omega C_T}{g_m} \frac{i_{nd}}{\sqrt{\Delta f}} \quad (4.8c)$$

As shown in Fig. 4.5b, when the input is open-ended, the noise generator $\overline{v_n^2}$ can be expressed as

$$\frac{v_n}{\sqrt{\Delta f}} = \frac{v_{n,1}}{\sqrt{\Delta f}} + \frac{v_{n,2}}{\sqrt{\Delta f}} \quad (4.9a)$$

where $\overline{v_{n,1}^2}$ is the input-referred noise from transistor M_1, and $\overline{v_{n,2}^2}$ is from L_S caused by $\overline{i_n^2}$. They are expressed as

$$\frac{v_{n,1}}{\sqrt{\Delta f}} = \frac{i_{nd}}{g_m \sqrt{\Delta f}} \quad (4.9b)$$

and

$$\frac{v_{n,2}}{\sqrt{\Delta f}} = j\omega L_S \frac{i_n}{\sqrt{\Delta f}} \quad (4.9c)$$

From (4.9a) to (4.9c), the equivalent noise voltage generator is obtained as

$$\frac{v_n}{\sqrt{\Delta f}} = \frac{i_{nd}}{g_m \sqrt{\Delta f}} + j\omega L_S \frac{i_n}{\sqrt{\Delta f}} \quad (4.9d)$$

Normally, the input-referred noise voltage source $\overline{v_n^2}$ is partially correlated with the input-referred noise current source $\overline{i_n^2}$. Hence, $\overline{v_n^2}$ can be expressed as the sum of two components, one fully correlated, $\overline{v_{n,c}^2}$, and the other, $\overline{v_{n,u}^2}$, uncorrelated to the noise current source $\overline{i_n^2}$ as

$$\frac{v_n^2}{\Delta f} = \frac{v_{n,c}^2}{\Delta f} + \frac{v_{n,u}^2}{\Delta f} \quad (4.10)$$

and the corresponding correlation impedance Z_c can be expressed as [3]

$$Z_c = \sqrt{\frac{\overline{v_{n,c}^2}}{\overline{i_n^2}}} = j\omega L_S + \frac{1}{j\omega C_T} \frac{1 + |c|p\alpha\chi}{1 + 2|c|p\alpha\chi + (p\alpha\chi)^2} \tag{4.11}$$

where [6]

$$c = \frac{\overline{i_{ng}i_{nd}^*}}{\sqrt{\overline{i_{ng}^2}\,\overline{i_{nd}^2}}} \approx -j0.395 \tag{4.12a}$$

$$p = \frac{C_{gs}}{C_T} \tag{4.12b}$$

$$\chi = \sqrt{\frac{\delta}{5\gamma}} \tag{4.12c}$$

$$\alpha = \frac{g_m}{g_{d0}} \tag{4.12d}$$

In the above formula, c is the noise correlation coefficient between the gate noise and the drain noise, α represents the short-channel effects and is used to estimate the transconductance reduction due to the velocity saturation and mobility decrease for vertical fields.

The other two important parameters in the NF calculation are the uncorrelated noise sources $\overline{i_n^2}$ and $\overline{v_{n,u}^2}$. The corresponding equivalent noise conductance G_n or resistance R_u can be expressed in the following equations [3]:

$$G_n = \frac{\frac{\overline{i_n^2}}{\Delta f}}{4KT} = \frac{\gamma}{\alpha^2 g_{d0}}(\omega C_T)^2 \left(1 + 2|c|p\alpha\chi + (p\alpha\chi)^2\right) \tag{4.13}$$

$$R_u = \frac{\frac{\overline{v_{n,u}^2}}{\Delta f}}{4KT} = \frac{\gamma}{\alpha^2 g_{d0}} \frac{(p\alpha\chi)^2 \left(1 - |c|^2\right)}{1 + 2|c|p\alpha\chi + (p\alpha\chi)^2} \tag{4.14}$$

Following the conventional two-port system noise analysis, the NF of the LNA can be written as [2]

$$F = 1 + \frac{R_u + |Z_c + Z_S|^2 G_n}{R_S} \tag{4.15}$$

where $Z_S = R_S + jX_S$ is the source impedance.

When the source impedance is selected as $Z_S = Z_{opt} = R_{opt} + jX_{opt}$, the minimum NF can be realized, and the corresponding Z_{opt} can be determined according to as [2]:

$$R_{opt} = \sqrt{\frac{R_u}{G_n} + R_c^2} = \frac{(p\alpha\chi)\sqrt{1 - |c|^2}}{\omega C_T \left(1 + 2|c|p\alpha\chi + (p\alpha\chi)^2\right)} \tag{4.16a}$$

and

$$X_{opt} = -X_c = -\omega L_S + \frac{1}{\omega C_T} \frac{1 + |c|p\alpha\chi}{1 + 2|c|p\alpha\chi + (p\alpha\chi)^2} \tag{4.16b}$$

From (4.12a)–(4.12d), $|c| = 0.395$, $p < 1$, $\alpha \le 1$, and $\chi < 1$, therefore the coefficient of $\frac{1}{\omega C_T}$ in (4.16b) is close to one, hence the optimum source impedance can be roughly achieved if the series combination of C_T and L_S can be resonated over the interested frequency band [3]. With the help of the three-section Chebyshev BPF input network, the overall input reactance looking into the filter is resonated over a wide bandwidth, so $X_{opt} = 0$ is generated over the wide bandwidth. As a result, a quasi-minimum NF can be achieved over the entire LNA's bandwidth and the corresponding NF can be simplified from (4.15) as [3]:

$$F \approx 1 + \frac{R_u}{R_S} + G_n R_S \tag{4.17}$$

With the help of above derived equivalent noise parameters, the final expression of NF can be derived by substituting (4.13)–(4.16a), (4.16b) into (4.17) as

$$F \approx 1 + \frac{\gamma}{\alpha g_m R_S} \left[\frac{(p\alpha\chi)^2 \left(1 - |c|^2\right)}{1 + 2|c|p\alpha\chi + (p\alpha\chi)^2} + (\omega C_T R_S)^2 \left(1 + 2|c|p\alpha\chi + (p\alpha\chi)^2\right) \right] \tag{4.18}$$

For CMOS transistors, the transconductance g_m can be derived from the following saturation drain current equation applicable for both long and short channel devices [2]:

$$I_D = \frac{\mu_n C_{ox}}{2} \frac{W}{L} (V_{gs} - V_t) [(V_{gs} - V_t) \| (L E_{sat})] = W L C_{ox} v_{sat} E_{sat} \frac{\rho^2}{1 + \rho} \tag{4.19}$$

where E_{sat} is the field strength at which the carrier velocity drops to half of the value extrapolated from the low-field mobility, V_t is the threshold voltage, and

$$v_{sat} = \frac{\mu_n}{2} E_{sat}, \quad \rho = \frac{V_{gs} - V_t}{L E_{sat}} = \frac{V_{od}}{L E_{sat}} \tag{4.20}$$

The transconductance can be obtained from (4.19) as

$$g_m \equiv \frac{\partial I_D}{\partial V_{gs}} = \frac{1 + \rho/2}{(1+\rho)^2} \left[\mu_n C_{ox} \frac{W}{L} V_{od} \right] = \alpha \left[\mu_n C_{ox} \frac{W}{L} V_{od} \right] = \alpha g_{d0} \qquad (4.21)$$

It is noted that $p < 1$, $\alpha \leq 1$, and $\chi < 1$ from (4.12a) to (4.12d). From (4.18), it is seen that a larger transconductance will produce a better noise performance. Also, as shown in (4.21), for fixed g_m, smaller size transistor would result in bigger α, and hence better NF. From $g_m = \sqrt{2\mu_n C_{ox} \frac{W}{L} I_D}$ [5], for fixed g_m, smaller transistor means larger I_D; therefore, larger bias current is preferred for NF performance.

During the noise analysis for the amplifying transistor M_1, the average NF over the entire operating frequency band is also an important parameter to evaluate because of the ultra-wide operating band of the LNA. For a specific bias current I_{bias}, there is a range for the width of M_1 that can be chosen to achieve the minimum average NF. Therefore, in the UWB LNA design, as long as the noise performance is mainly limited by the contribution of M_1, which is the case as for the designed UWB LNA employing the cascoded topology, better noise performance of the LNA can be achieved if larger bias current is applied as deduced from (4.18).

4.2.4   UWB LNA Design

In the UWB LNA design, the bias current $I_{bias} = 5\,\text{mA}$ is assumed, and the minimum length of 0.18 μm is selected for both transistors M_1 and M_2. Considering the balance between the thermal drain noise and induced gate noise, the size of the amplifying transistor M_1 is selected as 260 μm using (4.18). While for the cascoded transistor M_2, in order to reduce the parasitic capacitances, a smaller size is preferred. Furthermore, a lower limit to the width of M_2 is set by its noise contribution because a smaller size transistor produces a higher noise [3]. Consequently, the final width of 60 μm is selected for M_2.

As the on-chip spiral inductors available in the Jazz 0.18 μm CMOS PDK used for the UWB LNA design only provide the rectangular structure with low quality factor, which is not suitable for the UWB LNA, it is necessary to design our own integrated spiral inductors with better Q and desired inductive values over the UWB operating band for the UWB LNA. To that end, EM simulations are performed based on the employed multi-layer CMOS structure. This inductor design will be described in the next section.

In the UWB LNA circuit analysis, the gate-drain capacitance of M_1 C_{gd} is first omitted for simplicity. However, at high frequencies, the presence of C_{gd} complicates the input impedance Z_{IN} and makes it differ from the assumed simple series RLC model. Therefore, during the actual circuit simulations, C_{gd} is included, so the values of the on-chip components are optimized through the circuit simulations. The final component parameters are presented in Table 4.2.

Table 4.2 Final component values of the UWB LNA

L_1 (nH)	C_1 (pF)	L_2 (nH)	C_2 (pF)	L_G (nH)	C_P (pF)	L_S (nH)	L_L (nH)	R_L (Ω)
1.08	0.65	1.63	0.45	1.32	0.06	0.61	2.65	85

Fig. 4.6 Source-follower buffer for the UWB LNA

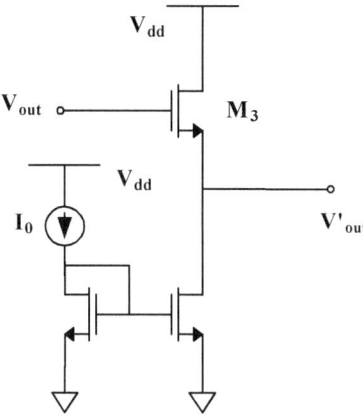

To facilitate on-wafer measurement for the UWB LNA, a typical source-follower buffer is also included in the LNA to drive an external 50 Ω load. This source-follower buffer is shown in Fig. 4.6, where a 50 Ω load is connected to the source of the transistor M3. The buffer consists of M_3 and a current mirror to provide the independent biased current source of 5 mA. The parameters (length and width) of the two transistors in the current mirror are optimized to produce higher output impedance. The size of M_3 is selected as 60 μm to achieve the transconductance of $g_{m3} = 1/R_{ext} = 1/50$ S. As the output voltage of the buffer is only half of that produced by the UWB LNA without buffer, the gain of the final UWB LNA structure with the buffer is 6 dB lower than that of the UWB LNA core. The performance of the designed UWB LNA will be presented in Sect. 4.2.6.

4.2.5   On-Chip Spiral Inductor Design

Typical RFICs employ on-chip MIM capacitors and spiral inductors for their matching networks instead of transmission lines, especially those operating in the low microwave frequency range. Typical on-chip spiral inductors have much lower Q than that of MIM capacitors and hence cause more detrimental effects to the RFIC performance. To that end, on-chip spiral inductors with decent Q are highly desired. Furthermore, for high operating frequencies, they should also have high self-resonant frequencies sufficiently far beyond the RFIC's operating frequency range. The Q of on-chip spiral inductors on silicon-based RFICs degrades substantially at high frequencies mainly due to the highly conductive (and hence lossy) silicon substrate [7]. Consequently, the design of high-Q on-chip spiral inductors is one of the major bottlenecks in CMOS RFIC design.

On-chip spiral inductors can be configured in various shapes, notably square, octagonal, and circular shapes. The design of on-chip spiral inductors can be found in details in [8]. On-chip spiral inductors can be realized in a single metal layer or multiple metal layers. Among all the possible configurations, the single-layer square spiral inductor is perhaps the most commonly used structure because of its relatively small area and ease in layout. Single-layer square spiral inductors are included in typical CMOS foundries' PDKs. Unfortunately, the Q of the square spiral inductor is not optimal as compared to other geometries. Another problem is that the noise coupling from the silicon substrate is potentially large due to the relatively large occupied die size and the small distance between the silicon substrate and spiral inductors. Therefore, selecting an optimal inductor structure and increasing the isolation between the inductor and silicon substrate would enhance the overall performance of the spiral inductor. Based on this consideration, octagonal-shape inductors with patterned ground shield (PGS) are designed to be used in the UWB LNA. This topology is fully compatible with CMOS technology and has the advantage of increased Q and isolation from the silicon substrate.

Various equivalent-circuit electrical models have been used for silicon on-chip spiral inductors. Figure 4.7 shows one of those models, referred to as the π-model. It consists of C_1, R_S, L, C_P, R_{sb}, and C_{sb} [2]. The series capacitance C_1 represents the capacitance due to the overlaps between the spiral and the center-tap underpass, which is determined by the space between the two metal sections. R_S is the resistance from the metal traces and is controlled by the sheet resistance and the length/width ratio of the metal traces. This resistance takes into account the losses due the skin effect in the spiral interconnect structure and the induced eddy current in conductive media (primary the silicon substrate) close to the inductor. C_P represents the parasitic oxide capacitance between the inductor and silicon substrate. R_{sb} and C_{sb} stand for the silicon substrate's parasitic resistance and capacitance, respectively.

Fig. 4.7 Inductor π-model

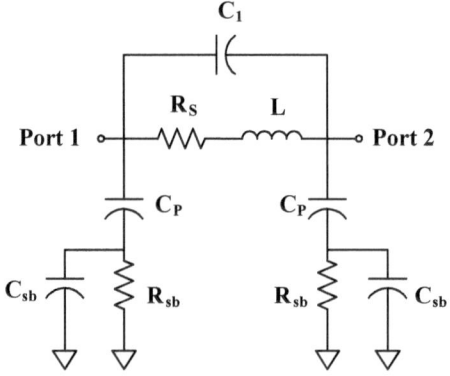

Since the resistance R_S of the inductor metal traces causes the primary energy loss, reducing it would increase the Q of the inductor. Additionally, reducing the substrate loss due to the magnetic and electrical couplings to the substrate from the inductor's metal traces further increases the Q of the inductor. To reduce the metal trace's resistance, a wide strip is usually used. Multi-layer metals can be used to reduce the inductor area or increase the inductance per area. However, using multiple metal layers has a disadvantage of generating parasitic capacitances between each layer that tend to reduce the inductor's self-resonant frequency [2].

From the model shown in Fig. 4.7, it can be seen that the inductor's parasitic effects of the substrate are a very important factor for inductor performance. To reduce the parasitic effects, a shield between inductor and substrate can be considered. For a solid ground shield, the inductor's magnetic field is disturbed and induces an eddy current in the solid ground shield, which flows in the opposite direction of the current in the spiral. This negative mutual coupling results in reduced inductance, hence making a solid ground shield less attractive. To increase the resistance to the eddy current, a patterned ground shield (PGS) with slots orthogonal to the spiral [9] can be used, resulting in reduced eddy current loss. The corresponding spiral structure is shown in Fig. 4.8, where the slots act as an open circuit to cut off the path of the induced eddy current.

To effectively cut off the eddy current, the slots should be narrow enough so that the vertical electric field emerging from the inductor cannot penetrate through the PSG into the underlying silicon substrate. To prevent the negative mutual coupling, the ground ring under the spiral inductor is intentionally broken into several sections to reduce the current loop effects [9]. To minimize the impedance to ground just like any ground paths or connections used in RF circuits, the ground ring should be grounded as close as possible to the actual ground.

Fig. 4.8 Layout of a PGS spiral inductor

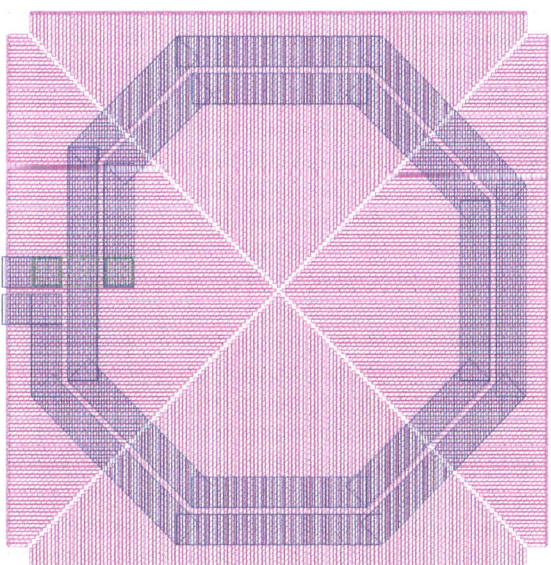

Based on the Jazz 0.18 μm RFCMOS process, the top metal layer M6 is selected to construct the octagonal spiral inductors. Comparing with the other five thinner metal layers, the resistance in the M6 metal traces is smaller. Metal layer M1 is selected to form the PGS. The spiral inductor's (top) metal traces and (bottom) PGS can be seen in Fig. 4.8. The EM software Zeland IE3D [10] is used to design and optimize the parameters of the top-layer spiral inductors as well as the PGS to achieve the desired Q and the inductance value over the operating frequency band for the spiral inductors of the designed UWB LNA (L_1, L_2, L_G, L_S, and L_L in Fig. 4.2). These inductors are also fabricated separately on the same UWB LNA chip for measurement to verify their design. To measure the S-parameters of the fabricated inductors, calibration components including open, short, through, and on-wafer RF pads are also fabricated and measurements are made on the same wafer to de-embed the RF pad effect. Two-port S parameters are measured for all the five fabricated inductors over 2–12 GHz using a vector network analyzer and RF probe station. The parameter extraction is performed by IE3D with the de-embedded S-parameters to produce the measured Q and inductance values over the entire frequency band.

Figure 4.9 shows the measured and simulated results for the designed spiral inductor L_1. The maximum Q of around 16 appears at 5.5 GHz and over the whole UWB band, it is larger than 10 as can be seen in Fig. 4.9a. From Fig. 4.9b, it is clear that the inductance is almost independent to the frequency variation and hence is useful for circuit design. Comparing to the inductor with the same inductance available in the PDK, the Q of the designed spiral inductor L_1 is indeed better. This better Q would lead to improved NF for the UWB LNA and hence enhanced performance. Similar performances are obtained for the remaining four spiral inductors designed for the UWB LNA.

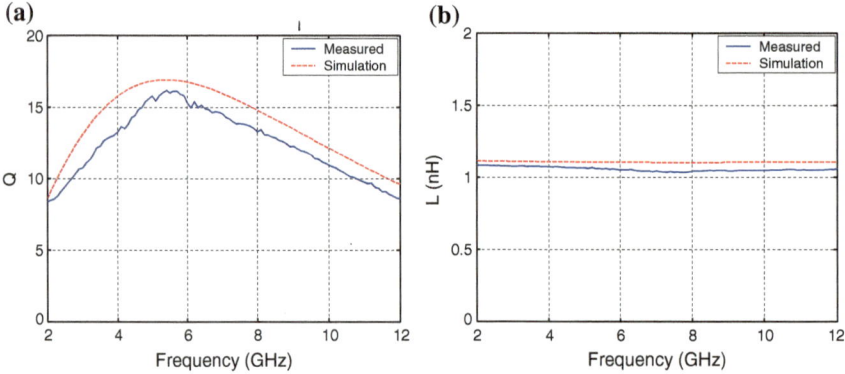

Fig. 4.9 Performance of the PGS spiral inductor L_1: **a** Q and **b** inductance L

4.2.6 UWB LNA Fabrication and Performance

Figure 4.10 show a photograph of the designed UWB LNA fabricated in Jazz
0.18 μm RFCMOS process. Its overall size is 0.88 mm × 0.7 mm including the
on-wafer RF and dc pads. The UWB LNA was measured on-wafer using a vector
network analyzer (for S-parameters), noise figure instrument (for NF), pulse generators and digitizing oscilloscope (for time-domain pulse response), and probe
station (for on-wafer probing). It is noted that the UWB LNA contains the buffer
integrated at the output of the core of the UWB LNA for measurement purposes.
This buffer will be removed from the UWB LNA when the LNA is integrated in the
final UWB receiver.

Figure 4.11 shows the measured and simulated return losses (S_{11}) at the input
port, which agree reasonably well. The measured input return loss is greater than
10 dB over the frequency range from 2.9 to 12 GHz.

Figure 4.12 presents the measured and simulated return losses at the output port
(S_{22}), which match very well. Over 2–12 GHz, the output return loss is more than
10 dB. This good return loss is achieved due to the use of the buffer at the output
the UWB LNA.

Figure 4.13 shows the measured and simulated results of the reverse isolation
(S_{12}) of the UWB LNA, which are very close to each other. Over the frequency
band of 2–12 GHz, more than 40 dB isolation is achieved, which proves the
effectiveness of the cascoded LNA structure. Figure 4.14 presents the gain (S_{21})
performance of the UWB LNA with the buffer stage. Over the 3 dB bandwidth of
2.6–9.8 GHz, a minimum power gain of 9.4 dB is achieved, which is made possible

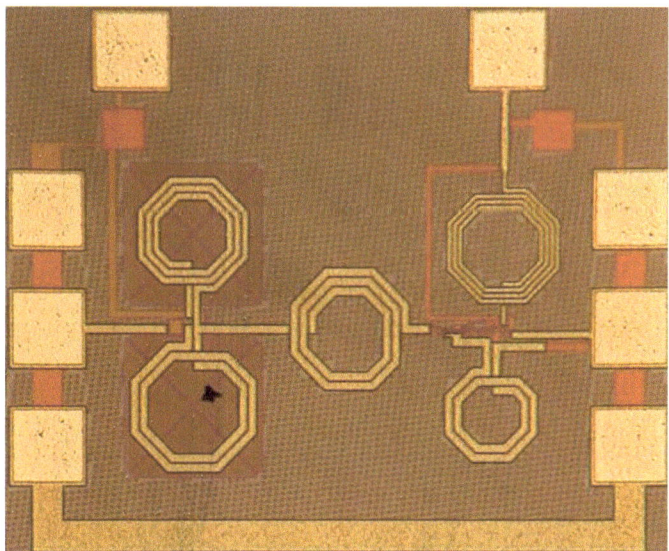

Fig. 4.10 Photograph of the UWB LNA chip

Fig. 4.11 Input return loss
(S_{11}) of the UWB LNA (with
buffer)

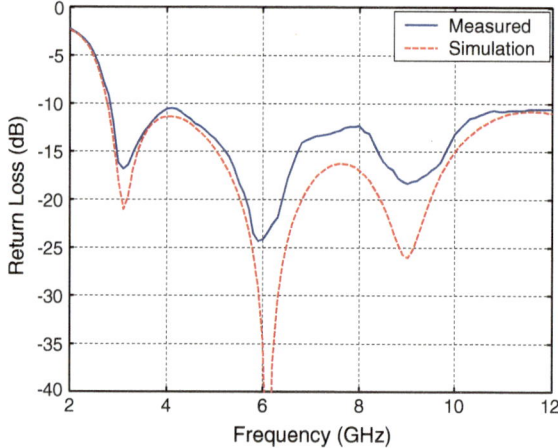

Fig. 4.12 Output return loss
(S_{22}) of the UWB LNA (with
buffer)

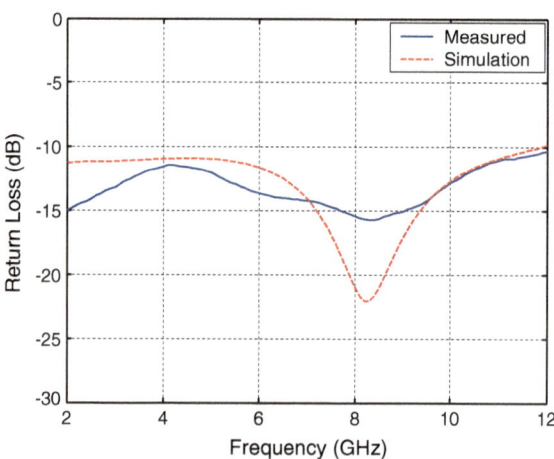

Fig. 4.13 Reverse isolation
(S_{12}) of the UWB LNA (with
buffer)

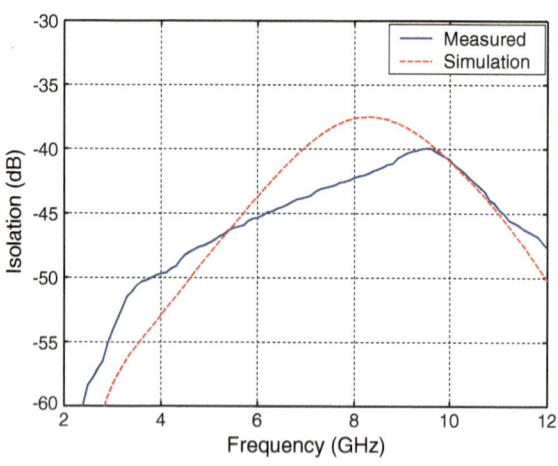

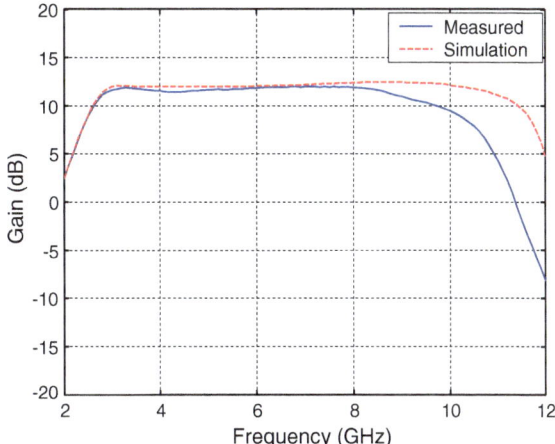

Fig. 4.14 Power gain (S_{21}) of the UWB LNA with buffer

with the use of the shunt-peaking topology. Over this frequency band, the maximum gain is 12.4 dB. Furthermore, the ripple of the gain is very small over the whole band. The difference between the measured and simulated results at the high-frequency end is primarily caused by the extra parasitic capacitance from the output buffer, which was not fully considered during the simulation.

Another important parameter for the UWB LNA design for (time-domain) pulse applications is the phase linearity of the LNA over the operating frequency band. To that end, the phase response (phase of S_{21}) of the UWB LNA was measured and presented along with that simulated in Fig. 4.15. The measured phase response shows very linear phase response over the entire UWB band, which also matches well with the simulation result. Linear phase response and, hence, constant group velocity across the operating frequencies is very critical for UWB pulse applications. The exhibited highly linear phase response helps produce the UWB pulse

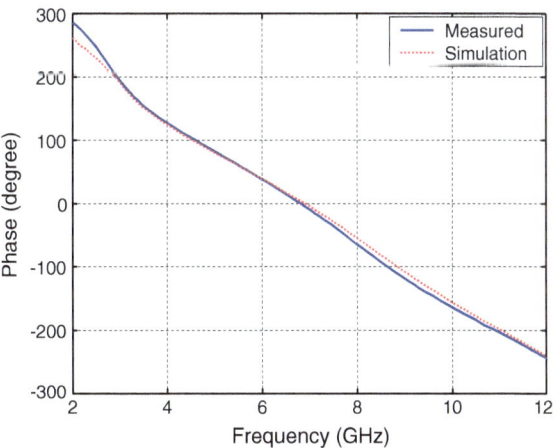

Fig. 4.15 Phase performance of S_{21} for the UWB LNA

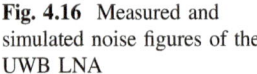

Fig. 4.16 Measured and simulated noise figures of the UWB LNA

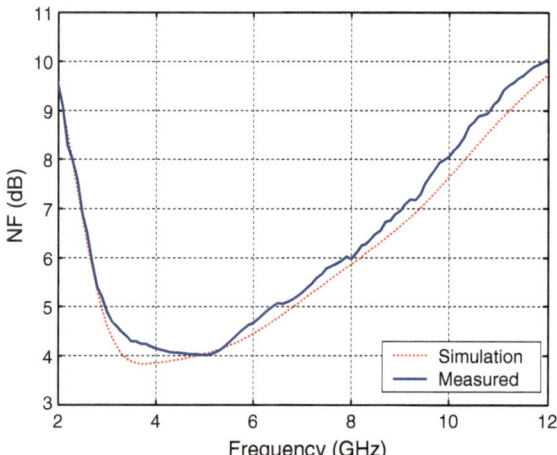

signals with less distortion at the UWB LNA's output that resemble well the UWB pulse signals fed to the LNA.

Figure 4.16 shows the noise performance over the entire UWB band. The measured minimum NF is 4 dB at 5.2 GHz, while the average NF over the entire UWB band is around 5.8 dB. As can be seen, the measured and simulated NF agree well.

The main different characteristic between a UWB LNA (or any RF component) operating in continuous wave (CW) frequency domain and that operating in pulse time domain (or simply impulse mode), even they have the same operating frequency range, is that the former processes sinusoidal signals sequentially in CW mode, while the latter processes (non-sinusoidal) pulse signals (i.e., it processes all the frequency components of a pulse signal simultaneously). The CW UWB LNA is thus basically operated over an ultra-wide bandwidth of single-frequency signals. On the other hand, the (non-sinusoidal) pulse UWB LNA is operated with a true ultra-wideband pulse signal having all frequencies across the ultra-wide band occurring at the same time. Therefore, the characterization of an UWB LNA for UWB impulse applications should also be carried out in the time domain besides the frequency domain.

To assess the performance of the UWB LNA and its suitability for UWB impulse applications, both simulation and measurement were conducted in the time domain. The time-domain measurement was performed on-wafer with a digitizing oscilloscope, pulse generators and on-wafer probe station. Figure 4.17 shows the simulated and measured signals that the UWB LNA produces from the input signal of a monocycle pulse having 3 dB pulse width of 150 ps and peak-to-peak amplitude of nearly 0.1 V. The measured output signal has the peak-to-peak voltage of 0.3 V with almost symmetric pulse shape and relatively small ripple. The measured output waveform also resembles closely that of the input signal, demonstrating that the designed UWB LNA can reproduce faithfully the waveform of an UWB non-sinusoidal pulse signal. This high signal fidelity is a critical

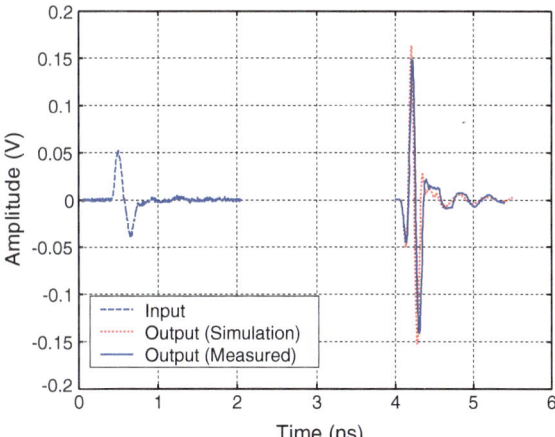

Fig. 4.17 Measured and simulated UWB LNA's pulse performance in time-domain

requirement for UWB impulse systems, as the UWB pulse signal carrying information must be transmitted with minimum distortion. The agreement between the simulated and measured pulses, as seen from Fig. 4.17, is also good. Compared to the simulated waveform, however, there is some pulse-width expansion in the measured waveform because of the UWB LNA's gain roll-off in the high-frequency region.

4.3   UWB Correlator Design

Comparing to typical narrow- and wide-band receivers operating in the frequency domain, the impulse UWB receiver has a much simpler architecture topology. The UWB impulse receiver can be formed with a simple topology consisting of a UWB LNA, a wideband correlator, and a high-frequency analog-to-digital converter (ADC) [6]. The basic function of the correlator is to convert RF signals received by the LNA to a baseband signal for detection. Figure 4.18 shows a typical correlator that consists of a multiplier followed by an integrator. The two inputs to the correlator, or specifically the multiplier, are the input monocycle pulse signal from the UWB LNA's output and its local template monocycle pulse signal either generated on the chip internally or supplied to the chip externally. The received monocycle pulse signal is correlated with the template monocycle pulse during a certain period, normally the pulse repetition period or several pulse periods used for one symbol, and the output signal from the correlator is sampled and held to detect whether there is a signal in the observing window.

In theory, a correlator can be implemented either in analog or digital format. However, for digital format, although a correlator can be realized for high data rates of up to 100 Mbps at high bandwidth, direct sampling of 3.1–10.6 GHz frequency signals is ultimately required, which is almost impossible for current ADC

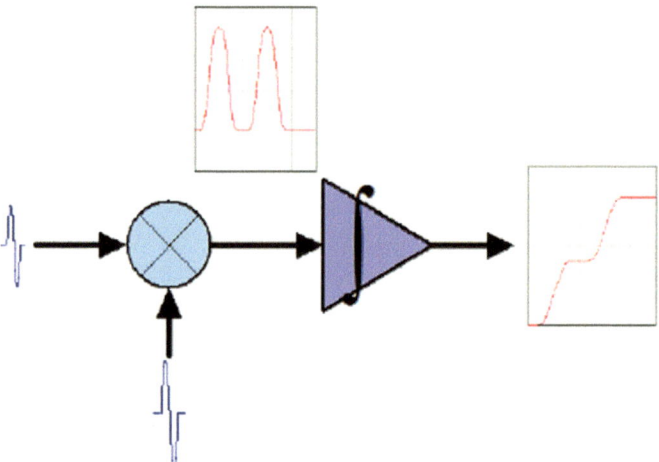

Fig. 4.18 Correlator in UWB receivers

techniques. Furthermore, a digital correlator normally consumes more power and is lower in efficiency when compared with its analog counterpart. Considering above factors, a analog correlator is preferred.

The advantage of an analog correlator is that it can process signals in real time and provide a continuous output at low frequency, thereby can remove the need of special requirements for the ADC in the receiver [11]. An analog correlator is therefore well suited for UWB receiver implementations, where its constituent analog multiplier is required to have a good linearity for low signal distortion.

As the correlator is used to detect the signal presence with known waveform in a noisy background, its output is zero for noise-only condition. In operation, the correlated input signal (to the correlator) is integrated with the local template signal over the pulse duration to produce an output signal with certain voltage. The cross-correlation function involved in this process can be expressed as:

$$f = \int_{t=t_0}^{t=t_0+T} RF(t)LO(t)dt \tag{4.22}$$

where $LO(t)$ is the local template signal, $RF(t)$ is the input RF signal of the correlator, and T is the integration period.

For the designed UWB receiver operating from 3.1 to 10.6 GHz, the multiplier of the correlator is required to have a wide bandwidth up to 10.6 GHz, which assures the output signal's waveform preserve the input pulse shape. This brings great challenge to the design of CMOS analog multipliers for such UWB receiver. Most of the published CMOS analog multipliers can only operate at low frequencies [12–15]. Although some UWB mixer designs can achieve very broad bandwidth [16, 17], information on the time-domain performance, which is an important consideration in impulse-type UWB receiver design, is not provided, making it rather difficult to assess their full potential for impulse UWB receivers.

In this section, a UWB four-quadrant multiplier is presented, which can be used for the UWB receiver's correlator.

4.3.1   DC Analysis

The schematic of the four-quadrant multiplier, which is based on the transconductor multiplier structure reported in [18], is shown in Fig. 4.19. The central component of this multiplier is the CMOS programmable transconductor. As a current-mode element, it converts an input voltage signal into a differential current to realize the multiplication.

The differential structure in Fig. 4.19 enables the even-order terms generated by the nonlinear components to be cancelled, therefore enhancing the linearity of the multiplier. In order to reduce the leakage of the input RF signal to the output port, a pair of NMOS transistors (M_9 and M_{10}) is inserted between the outputs of the transconductor M_5–M_8 and the multiplier's output. To compensate for the gain roll-off at high frequencies, the shunt-peaking topology is employed, just like the case in the UWB LNA design. Two inductors L_1 and L_2 with optimized values are added in series with the load resistors at the output (drain terminals of M_9 and M_{10}), hence improving the gain performance at the high-frequency end was improved and achieving the wide bandwidth. Moreover, two source-follower buffers are also included to facilitate the intermediate frequency (IF) signal measurement.

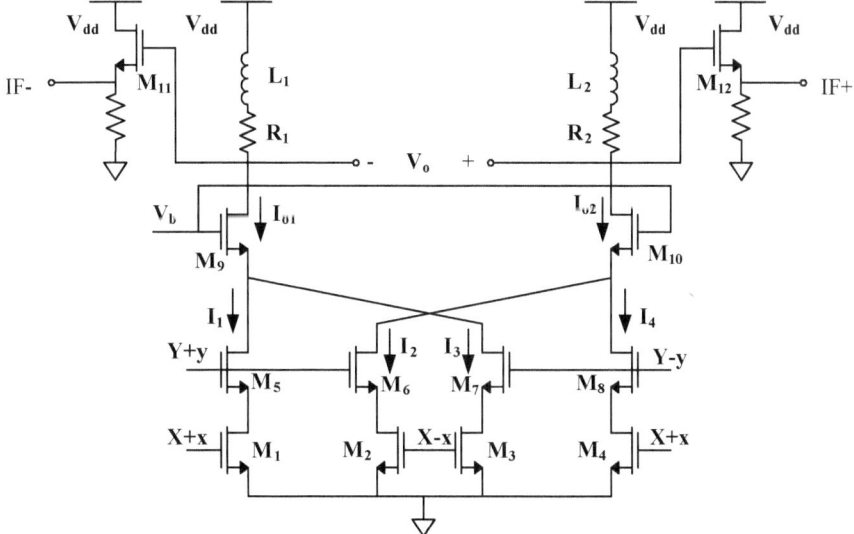

Fig. 4.19 Schematic of the four-quadrant multiplier

As shown in Fig. 4.19, the RF signal "x" (shown in $X + x$) enters the lower branches formed by the transistors M_1–M_4, which operate in the linear region through a bias voltage of X. On the other hand, the transistors M_5–M_8, used as the signal path for the LO template signal y (shown in $Y + y$) operate in the saturation region when a proper dc bias voltage Y is supplied. Under the condition of the triode region, the large-signal model of MOS transistors is applied to M_1–M_4, and the corresponding current flowing through each of the lower branches, I_1 to I_4, can be expressed as [19]

$$I_i = K\left(X \pm x - V_{tn} - \frac{V_{dsi}}{2}\right)V_{dsi} \tag{4.23}$$

where $K = \mu_n C_{ox} \frac{W}{L}$, V_{tn} is the NMOS threshold voltage, and V_{dsi} is the drain-source voltage of the i-th MOS transistor.

For the MOS transistors, the value of the transconductance g_m is dependent on the dc bias conditions. As g_m of the transistor in the saturation region is much larger than that of the transistor in the triode region, the upper transistors (M_5–M_8) operating in the saturation region can be considered as the source followers. Therefore, for the lower branch transistors M_1–M_4, the corresponding drain-source voltage V_{dsi} can be expressed as [19]

$$V_{dsi} = V_{ds} + y \tag{4.24}$$

for M_1 and M_2, and

$$V_{dsj} = V_{ds} - y \tag{4.25}$$

for M_3 and M_4. V_{ds} is the drain-source voltage of the transistor at the bias point under the condition of $x = y = 0$. As shown in Fig. 4.19, the total output current I_o can be derived as [20]

$$I_o = I_{o1} - I_{o2} = (I_1 + I_3) - (I_2 + I_4) = (I_1 - I_2) + (I_3 - I_4) \tag{4.26}$$

where

$$I_1 = K\left(X + x - V_{tn} - \frac{V_{ds} + y}{2}\right)(V_{ds} + y) \tag{4.27a}$$

$$I_2 = K\left(X - x - V_{tn} - \frac{V_{ds} + y}{2}\right)(V_{ds} + y) \tag{4.27b}$$

$$I_3 = K\left(X - x - V_{tn} - \frac{V_{ds} - y}{2}\right)(V_{ds} - y) \tag{4.27c}$$

$$I_4 = K\left(X + x - V_{tn} - \frac{V_{ds} - y}{2}\right)(V_{ds} - y) \qquad (4.27\text{d})$$

Hence, the final output current is obtained as

$$I_o = 2Kx(V_{ds} + y) - 2Kx(V_{ds} - y) = 4Kxy \qquad (4.28)$$

where we assume all the transistors (M_1–M_4) have equal size, hence same K for all transistors. The multiplication function is thus achieved, and the corresponding output voltage of the multiplier can be expressed as [19]

$$V_o = -I_o Z_o = -4KxyZ_o \qquad (4.29)$$

where Z_o is the output load between the $+$ and $-$ output.

4.3.2 AC Analysis

Figure 4.20 presents a simplified small-signal equivalent circuit used for the bandwidth analysis of the multiplier. Each of the transconductors (M_1 to M_8) is assumed to be an ideal current source with the parasitic capacitance of C at its output. Furthermore, to simplify the circuit analysis, the output resistance is omitted because of its much larger value compared with the impedance seen from the source of the cascoded transistors such as M_9 and M_{10}. In Fig. 4.20, L and R_L are the load inductor and load resistor (corresponding to L_1, L_2 and R_1,R_2 shown in Fig. 4.19), g_m and r_{ds} are the transconductance and output resistance of M_9 or M_{10}, and v_{gs} is the gate-source voltage of transistor M_9 or M_{10}. As shown in Fig. 4.19, in order to improve the bandwidth, the shunt-peaking topology is employed, where the output load resistance R_L is in series with the inductor L to compensate for the gain roll-off at the high-frequency end. For the output, under the condition of open circuit, the parasitic capacitance at the output can be ignored in the analysis because of its very small value. Typically, the drain node is the dominant pole, because the equivalent

Fig. 4.20 Simplified small-signal equivalent circuit for bandwidth analysis

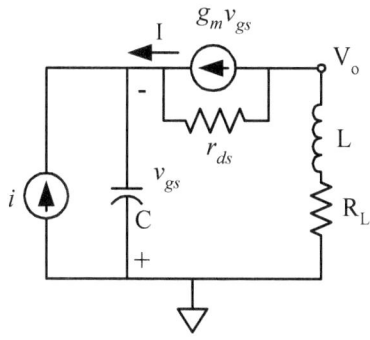

resistance seen at the source node, approximately $1/g_m$, is low. However, in the case considered here, the parasitic capacitance associated with the source node could be large because all three transistors are connected to the same node, thus producing the dominant pole [19].

The transfer function can be derived, from Fig. 4.20, as [19]

$$\frac{v_o}{i} = \frac{R_L + j\omega L}{\frac{-\omega^2 CL}{1 + g_m r_{ds}} + \frac{j\omega C(R_L + r_{ds})}{1 + g_m r_{ds}} + 1} \tag{4.30}$$

Hence, the dominant and non-dominant poles can be expressed, assuming the two poles can be separated from each other [19], as

$$\omega_{p1} = \frac{1 + g_m r_{ds}}{(R_L + r_{ds})C} \tag{4.31a}$$

$$\omega_{p2} = \frac{R_L + r_{ds}}{L} \tag{4.31b}$$

respectively, and the zero is $\omega_z = R_L/L$. According to the schematic simulation, ω_{p1} is around 2.2 GHz, while ω_{p2} is much higher than ω_{p1}. Hence ω_{p1} is the dominant pole for this case. In theory, under the condition that the dominant pole ω_{p1} can be cancelled by the zero ω_z [9], the bandwidth of the multiplier can be increased dramatically, hence the shunt-peaking topology effectively improves the bandwidth performance [2].

4.3.3   UWB Correlator Fabrication and Performance

The designed multiplier was fabricated with Jazz 0.18 μm CMOS process. The layout structure was arranged symmetrically to reduce the potential unbalance caused by non-symmetric structures. The on-chip octagonal-shape spiral inductors L_1 and L_2 were designed and optimized through EM simulations to achieve constant inductance over the frequency range from 3.1 to 10.6 GHz.

To calculate the frequency response, the RF and LO ports were fixed to dc and the input signal was directly fed to the source node of M_9 and M_{10}. First, the frequency response of the multiplier without the shunt-peaking inductor and load capacitor at the output was investigated, and the result is shown in Fig. 4.21. The 3 dB bandwidth is around 2 GHz. For comparison purpose, an additional 100 fF capacitance was connected to the same source node of M_9 (M_{10}), as shown in Fig. 4.19, and under this condition, the bandwidth was reduced to about 1 GHz as can be seen in Fig. 4.21, which validates the dominant-pole assumption. Figure 4.22 compares the frequency responses with and without the output buffer when the shunt-peaking inductors are used. It is obvious that, after the inductor of around 30 nH is included, the bandwidth is increased to 10 GHz, indicating that the

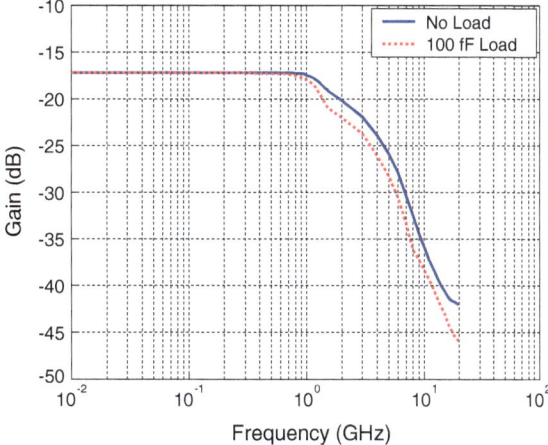

Fig. 4.21 Frequency response for the dominant pole

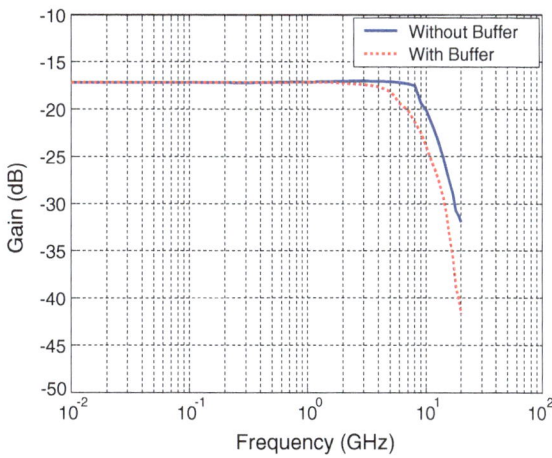

Fig. 4.22 Frequency response with and without the buffer considering the shunt-peaking inductor effect

pole-zero cancellation topology is indeed effective. In the case where the buffer is included at the output of the multiplier, the simulation result indicates that the bandwidth of the multiplier is reduced to 7 GHz, which is expected due to the extra capacitive loading from the buffer.

Figure 4.23 shows the fabricated multiplier chip having the size of 1 mm × 0.7 mm including the RF and dc bias pads. The measurement was conducted with RF probes used for the on-wafer RF and LO ports, while the IF ports were accessed off-chip through a package.

The measured conversion gain and RF-port return loss are shown in Fig. 4.24, where the IF frequency is fixed to 10 MHz and the LO power is −1 dBm. The measured conversion gain of the multiplier including the output buffer is more than 7 dB from 3 to 10 GHz. The agreement between the measured and simulated conversion gains is considered reasonably; the difference is mainly caused by the

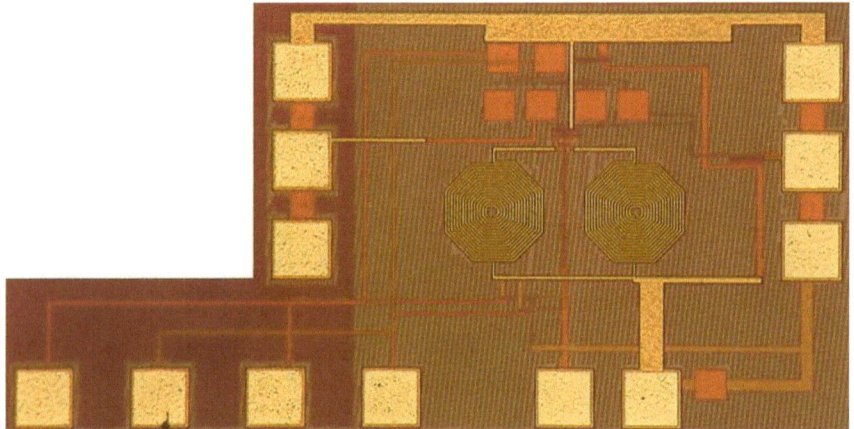

Fig. 4.23 Photograph of the fabricated multiplier

Fig. 4.24 Conversion gain and RF return loss with 10 MHz IF frequency, −1 dBm LO power, and −20 dBm RF power

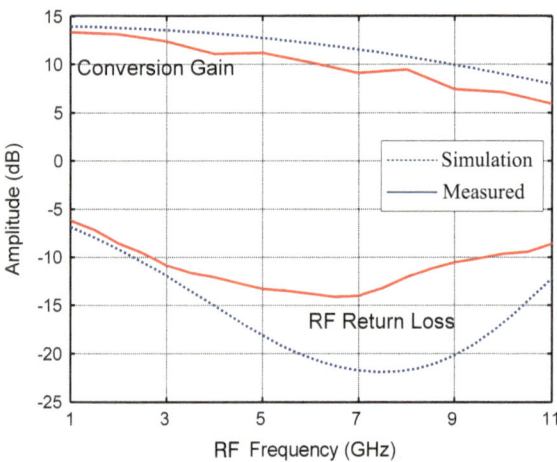

parasitic resistance loss from the shunt-peaking inductor and buffer, and the buffer's parasitic capacitor. More than 10 dB return loss at the RF port is measured across 3–10 GHz, which is reasonably matched to the simulated result.

4.4   UWB Receiver

The UWB receiver, as shown in Fig. 4.25, is formed by integrating the template tunable pulse generator, UWB LNA, and multiplier designed and described in the previous sections.

Fig. 4.25 Block diagram of the UWB receiver

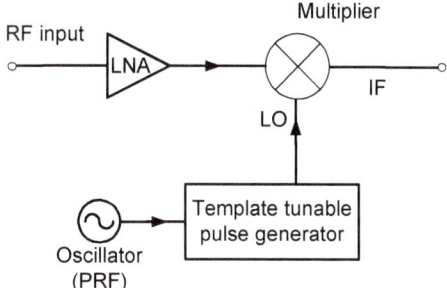

To simulate the time-domain response of the UWB impulse receiver, two monocycle pulses having the same width of 0.2 ns different amplitudes are used. The pulse with the smaller amplitude is applied to the RF input of the multiplier through the UWB LNA, while the larger pulse is fed to the LO port of the multiplier from the template tunable pulse generator. The IF signal is then generated at the output of the source-follower buffer. Figure 4.26 shows the simulated IF output signal in the time domain, which depends on the polarity of the received RF signal. When the RF pulse is in-phase with the LO pulse, the IF signal is positive, as seen in Fig. 4.26a. On the other hand, when the RF pulse is out-of-phase with LO pulse, the IF signal is negative, which is shown in Fig. 4.26b. These expected output signals are obtained at the output of the multiplier. They show that the multiplier has sufficient bandwidth and is able to work with the input subnano-second pulses. Figure 4.27 shows the layout of the UWB receiver including the template tunable pulse generator, UWB LNA, multiplier, and RF and dc pads, which occupies 1.4 mm × 0.7 mm die size.

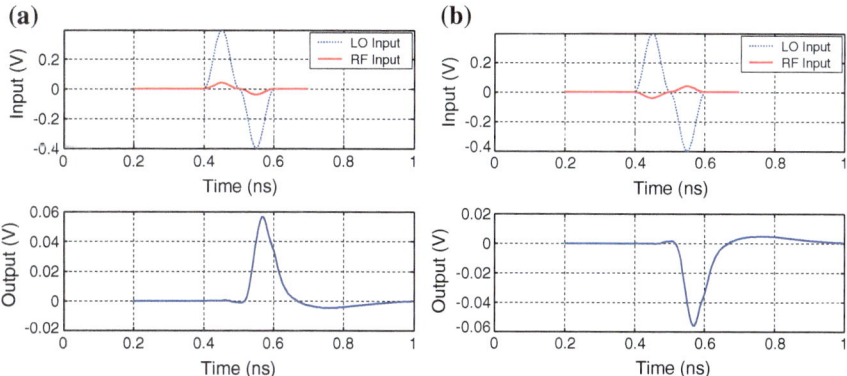

Fig. 4.26 Transient simulation of the UWB receiver with **a** RF and LO pulses in-phase and **b** RF and LO pulses out-of-phase

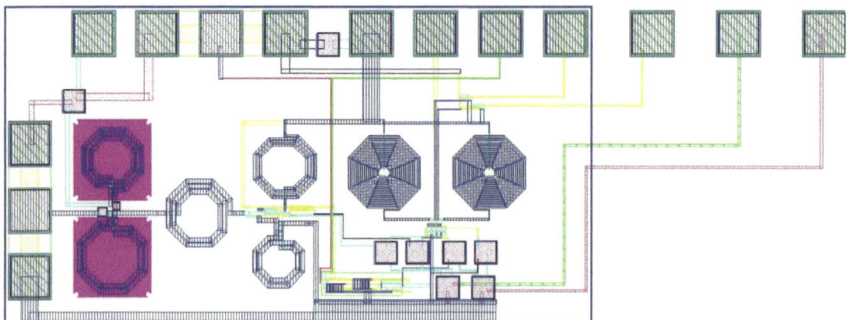

Fig. 4.27 Layout of the UWB receiver

4.5   Summary

This chapter presents the design of a UWB impulse receiver and its core components of UWB LNA and correlator. The UWB LNA employs the cascoded common-source inductively degenerated topology with an extended UWB ladder matching network. The shunt-peaking topology is also implemented for the LNA to further improve its performance at the high-frequency end. The LNA utilizes the PGS spiral inductors designed for high Q to further improve the LNA's performance. The return losses of the LNA with the source-follower buffer for both the input port and output port are better than 10 dB over the entire UWB band of 3.1–10.6 GHz. The reverse isolation of the LNA of more than 40 dB is also achieved over the UWB range. The UWB LNA achieves the maximum gain of 12.4 dB over the UWB band. The UWB LNA possesses an average NF of 5.8 dB over the entire UWB band. The UWB multiplier of the UWB correlator is based on the transconductor multiplier structure with the central component of CMOS programmable transconductors. It converts the input voltage signal into differential current to realize the multiplication. The shunt-peaking topology is also applied at the output, which achieves the pole-zero cancellation and extends the multiplier bandwidth from 2 to 10 GHz for un-load situation, and 7 GHz for buffer-load condition. Finally, the UWB LNA is integrated with the UWB multiplier and template pulse generator to form the UWB impulse receiver, which demonstrates its ability to receive and convert sub-nano-second pulse signals to baseband signals for UWB impulse systems.

References

1. M. Miao, C. Nguyen, Integrated CMOS impulse UWB receiver front-end design. Microw. Opt. Technol. Lett. **51**, 2590–2595 (2009)
2. T.H. Lee, *The Design of CMOS Radio-Frequency Integrated Circuits*, 1st edn. (Cambridge University Press, New York, NY, 1998)

3. A. Bevilacqua, A.M. Niknejad, An ultrawideband CMOS low-noise amplifier for 3.1–10.6-GHz wireless receivers. IEEE J. Solid-State Circuits **39**, 2259–2268 (2004)
4. *Advanced Design System* (Agilent Technologies Inc., Santa Clara, CA, 2006)
5. D.D. Wentzloff, A.P. Chandrakasan, Gaussian pulse generators for subbanded ultra-wideband transmitters. IEEE Trans. Microw. Theory Tech. **54**, 1647–1655 (2006)
6. N. Daniele, M. Pezzin, S. Derivaz, J. Keignart, P. Rouzet, Principle and motivations of UWB technology for high data rate WPAN applications, in *Proceedings of Smart Objects Conference,* Grenoble, France (2003)
7. Y. Zheng, H. Dong, Y.P. Xu, A novel CMOS/BiCMOS UWB pulse generator and modulator, in *IEEE International Microwave Symposium Digest*, June 2004, pp. 1269–1272
8. C. Nguyen, *Radio-Frequency Integrated-Circuit Engineering* (John-Wiley & Sons, New York, 2015)
9. C.P. Yue, S.S. Wong, On-chip spiral inductors with patterned ground shields for Si-based RF IC's. IEEE J. Solid-State Circuits **33**, 743–752 (1998)
10. *IE3D* (Zeland Software Inc., Fremont, CA, 2006)
11. C. Tu, B. Liu, H. Chen, An analog correlator for ultra-wideband receivers. EURASIP J. Appl. Signal Proc. **2005**(3), 455–461 (2005)
12. A.L. Coban, P.E. Allen, A 1.5 V four-quadrant analog multiplier, in *IEEE Circuits and Systems, Proceedings of the 37th Midwest Symposium*, vol. 1, Aug 1994, pp. 117–120
13. K. Tanno, O. Ishizuka, Z. Tang, Four-quadrant CMOS current-mode multiplier independent of device parameters. IEEE Trans. Circuits Syst. II: Analog Digit. Signal Process. **47**, 473–477 (2000)
14. G.A. Hadgis, P.R. Mukund, A novel CMOS monolithic analog multiplier with wide input dynamic range, in *IEEE Proceedings of 8th International Conference on VLSI Design*, Jan 1995, pp. 310–314
15. G. Colli, F. Montecchi, Low voltage low power CMOS four-quadrant analog multiplier for neural network applications. IEEE Proc. Int. Symp. Circuits Syst. **1**, 496–499 (1996)
16. M. Tsai, H. Wang, A 0.3—25-GHz ultra-wideband mixer using commercial 0.18-μm CMOS technology. IEEE Microwave Wirel. Compon. Lett. **14**, 522–524 (2004)
17. A. Motieifar, Z.A. Pour, G. Bridges, C. Shafai, L. Shafai, An ultra wideband (UWB) mixer with 0.18 μm RF CMOS technology. IEEE CCECE/CCGEI **14**, 522–524 (2004)
18. G. Han, E. Sanchez-Sinencio, CMOS transconductance multiplier: a tutorial. IEEE Trans. Circuits Syst. II: Analog Digit. Signal Process. **45**, 1550–1563 (1998)
19. L. Zhou, Y.P. Xu, F. Lin, A gigahertz wideband CMOS multiplier for UWB transceiver. IEEE Int. Symp. Circuits Syst. **5**, 5087–5090 (2005)
20. J.A.N. Noronha, T. Bielawa, C.R. Anderson, D.G. Sweeney, S. Licul, W.A. Davis, Designing antennas for UWB systems, in *Microwave & RF*, Jun 2003

Chapter 5
UWB Uniplanar Antenna

5.1 Introduction

UWB impulse systems communication and radar systems use ultra-short duration pulses (impulse or monocycle pulses as addressed here) in the sub-nanosecond regime, instead of the more conventional continuous sinusoidal waves, for transmission and reception of information. The pulse is of non-sinusoidal type and directly generates a very wide-band instantaneous signal having all frequencies occurring at the same time with various duty cycles depending on specific usages. This is very different from a sinusoidal signal that carries only a single frequency even it is used in UWB systems. A UWB impulse system works over an ultra-wide bandwidth at all times, while a UWB system employing sinuoidal signals only operates at a single frequency at a time. UWB impulse systems involve propagations of non-sinusoidal pulse signals through circuits, antennas, and air or other media. Effectively, UWB impulse systems transmit and receive all frequency components across extremely wide bandwidths simultaneously—not consequently as in UWB CW systems. This results in another constraint for the antenna design besides an ultra-wide bandwidth; in which the antenna needs to radiate or receive a short pulse signal without undesirable distortion on the transmitting or receiving signal waveform; that is radiating or receiving all the frequency components simultaneously with linear phase response over the entire ultra-wide frequency band. This high signal fidelity is a critical requirement for antennas used in UWB impulse communication and radar systems, as the UWB signals carrying information must be transmitted and received with minimum distortion. Compared to antennas used in CW systems, which only radiates or receives a CW signal at a single frequency each time, regardless of the operating bandwidth, and hence would easily maintain the transmitting or receiving waveform at the design frequency, antennas in UWB impulse systems is much more prone to alter the transmitting or receiving waveform and hence plays a much more important role in the signal analysis and formation of UWB impulse systems. UWB antennas are indeed the

© The Author(s) 2017   89
C. Nguyen and M. Miao, *Design of CMOS RFIC Ultra-Wideband Impulse Transmitters and Receivers*, SpringerBriefs in Electrical and Computer Engineering, DOI 10.1007/978-3-319-53107-6_5

key element dictating the transmitted and received pulse shape and amplitude in both time and frequency domain. They should have good impulse response with minimal distortion. The antenna gain is also important for UWB applications. The high-gain antenna is of benefit in respect of signal transmission efficiency and signal detection. In respect of transmission efficiency, the high gain antenna enables signal transmission with low loss of transmitting power. In respect of signal detection or reception, the high gain antenna enables reduction of the clutter signal coming from scatters other than the desired object, since the narrow beam width of the high gain antenna supports greater focusing of the system to the object. The antenna input reflection should be minimized over the entire antenna structure—not only at the input port but also at the antenna aperture and in between them—to avoid multiple reflections as the pulse travels along the antenna itself and between the antenna and transmitter or antenna and receiver, which may produce some clutter-like signals and increase the false alarm rate or mis-communications. Ideally, the performance of a UWB antenna including gain, radiation pattern, and matching should be constant across the entire operating band.

Many ultra-wide band antennas exist. However, a majority of these antennas does not meet the critical requirement of fidelity in signal transmission and reception, hence is not suitable for UWB impulse applications. One such example is the log-periodic antenna. While it is a frequency-independent antenna that can operate over extremely wide bandwidths, it is dispersive. Its smallest part radiates the highest frequency component while the largest part radiates the lowest frequency component. The log-periodic antenna hence radiates a chirp-like, dispersive waveform with different shapes at different azimuthal angles around the antenna because the dispersion-variation dependence on different direction and ranges.

There are several types of UWB antennas that can be used for short-pulse transmission and reception, such as TEM horn, bow-tie, Vivaldi and conical antennas [1–6]. Various impulse systems have used TEM horn type antennas because of their high gain, linear phase characteristic and easy fabrication. A modified version of the TEM horn antenna, called microstrip quasi-horn antenna, was developed [7] and used successfully in subsurface penetrating radars [8, 9]. Its time-domain performance demonststrates its suitability and usefulness for UWB impulse applications [10]. The microstrip quasi-horn antenna has seceral advantages as compared to the conventional TEM horn antenna. It uses a non-uniform transmission line structure realized with microstrip line, making it easy to connect with integrated circuits without any special transition. Its performance is similar to that of the TEM horn antenna, which has high gain and linear phase characteristic. In addition, when used as a receiving antenna, the microstrip quasi-horn antenna produces an output signal having a voltage waveform very similar to the incident electric field in the time domain, making it a preferred receiving antenna for making direct measurements of transient EM fields. Because of this characteristic, the microstrip quasi-horn antenna will be used later as the receiving antenna in a measurement system to verify the time-domain performance of the designed UWB planar antenna. Moreover, its physical size is relatively smaller than the TEM horn. Lastly, when two identical antennas are used for a bi-static configuration of a system, two microstrip quasi-horn antennas can be assembled into an antenna system with ease. For most commercial applications, however, the large size of the above-mentioned UWB antennas, including the microstrip quasi-horn structure, makes them not suitable for portable or handheld devices, or even in non-portable systems deployed in limited spaces. A few high-quality non-dispersive UWB antennas are also commercially available [7]. However, they have relatively large sizes.

To facilitate integration with transmitters and receivers employing RF integrated circuits and to suit portable or handheld systems, compact UWB antennas with uniplanar structures are preferred. Additionally, a "center-fed" uniplanar structure should be avoided due to the reason that the feed region of this structure lies in the heart of the most intense near-fields surrounding the antenna. Strong coupling between the feed structure and antenna seriously affects the near field and distorts the antenna radiation pattern [11].

In this chapter, we describe a low-cost, compact, easy-to-manufacture uniplanar UWB antenna that is omni-directional, radiation-efficient, and has a stable minimum-distortion impulse UWB response, which can be easily integrated with the designed UWB CMOS transmitter and receiver described earlier.

A remark needs to be mentioned here concerning the design, analysis and measurement of UWB antennas for impulse applications. UWB antennas transmit and receive (non-sinusoidal) pulse signals, and hence operating completely in the time domain. Although, from the Fourier series point of view, the frequency and time domains are correlated, and one can then view them as equivalent, they should be distinguished from one to another for UWB impulse applications. The design, analysis and measurement of a UWB antenna (or any other components) used in UWB impulse systems should be carried out in both frequency and time domains. In fact, due to its time-domain operation for impulse UWB applications, the time-domain performance of a UWB antenna is much more critical than its frequency counterpart. Furthermore, conducting antenna analysis in the frequency domain requires calculations of far-field values for both amplitude and phase, and these must be done over a very wide frequency range to be able to accurately extract the shape of the radiated pulse for UWB antennas, which make certain calculations very time consuming or, due to lack of computer memory, impossible to perform in some cases.

5.2 UWB Uniplanar Antenna Design

Figure 5.1 shows the UWB uniplanar antenna to be used with the UWB transmitter and receiver described in Chaps. 3 and 4, respectively. This antenna overcomes potential problems caused by the unwanted coupling between the fields from the feed network and antenna, as mentioned previously, and meets the requirements for UWB impulse applications, such as ultra-wide bandwidth, reasonable efficiency, satisfactory radiation properties, and linear phase characteristics. The operation

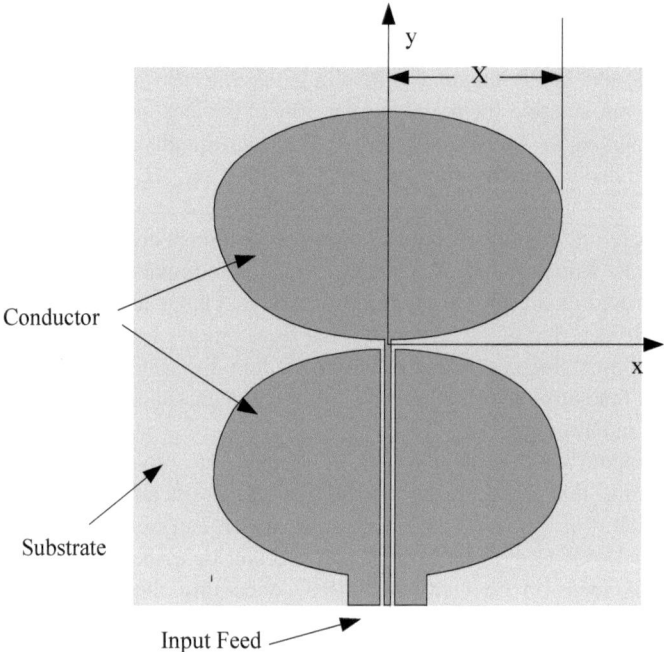

Fig. 5.1 UWB uniplanar antenna

of the antenna is based on the principles of (tapered) non-uniform transmission lines and traveling-wave antennas. This UWB uniplanar antenna can be considered as a uniplanar variant of the microstrip quasi-horn antenna [7] and TEM horn antenna [12] and hence can employ similar design procedures.

The UWB uniplanar antenna, as depicted in Fig. 5.1, consists of a two parallel slot lines with integrated coplanar waveguide (CPW) in the left part of the two slot lines. The CPW section from the SMA connector up to the left edge of the slot lines is not a part of the antenna; it is only used as a feed line, feeding signals to the antenna. Most of the signal's energy is confined within the CPW integrated within the slot lines, which is the CPW with (tapered) non-uniform ground planes, until it reaches the antenna center, where the signal is coupled from the CPW to the two parallel non-uniform slot lines and radiated out.

Assume the input (antenna center) and output (antenna's open end or antenna aperture) are matched to the corresponding source (50 Ω) and load (377 Ω) impedances at $x = 0$ and $\pm X$, as shown in Fig. 5.1, respectively, we can have [12]

$$\frac{dZ_0}{dx} = 0 \tag{5.1}$$

at the input ($x = 0$) and the output section ($x = \pm X$), where Z_0 is the characteristic impedance of non-uniform slot lines along the x direction. For a given maximum allowable input reflection coefficient $|R(0)|_{max}$, the optimum characteristic impedance variations of the NUTL can be obtained according to [12]

$$\log Z_0(x) = \frac{1}{2}\log[Z_0(0)Z_0(X)] + \frac{1}{2}\log\left[\frac{Z_0(X)}{Z_0(0)}\right]G\left(B,\frac{x}{X}\right) \qquad (5.2)$$

where X is the length of the non-uniform slot line as shown in Fig. 5.1, and $G\left(B,\frac{x}{X}\right)$ and its parameter B are given in [13]. The maximum allowable input reflection coefficient is

$$|R(0)|_{max} = \tanh\left[\frac{B}{\sinh B}(0.21723)\log\sqrt{\frac{Z_0(X)}{Z_0(0)}}\right] \qquad (5.3)$$

which shows that, for given input reflection coefficient $|R(0)|_{max}$ and the slot line's characteristic impedances at the antenna's input and aperture, the antenna length X is determined by the parameter B. For the special case of $B = 0$, the contour of the non-uniform slot lines, which dictates the antenna structure, is exponential taper.

The UWB uniplanar antenna as shown in Fig. 5.1 was fabricated on a Duroid substrate having 0.025-in thickness and relative dielectric constant of 10.5. The CPW feed line and the CPW integrated within the slot lines has the usual 50-Ω characteristic impedance. The characteristic impedance of the two parallel slot lines is 100 Ω. It is noted that the transition of from the 50-Ω CPW to two parallel 100-Ω slot lines should be as smooth as possible to minimize internal reflections. To that end, the slot-width of the slot lines should be close to the gap-width of the CPW. It is noted that the designed UWB uniplanar antenna is not integrated directly with the UWB transmitter and receiver on the same chip. Rather, it will be connected to a CMOS RFIC chip that contains the UWB transmitter and receiver for testing purposes. Therefore, the design of the 50-Ω CPW feed line need to consider the package specifications associated with the CMOS RFIC chip. In our case, a standard 52-lead LQFP open–package is used to accommodate the CMOS transmitter and receiver chip, which has 13 leads along each side with lead width of 0.012 in. and gap width between the leads of 0.014 in. [14]. To achieve a smooth transition from the signal and ground leads of the package accommodating the CMOS transmitter and receiver to the 50-Ω CPW feed line, the strip and gap widths of the 50-Ω CPW are chosen as0.020 in. and 0.010 in., respectively.

Exponentially tapered slot lines are chosen for the UWB uniplanar antenna to simulate an impedance transformer from 100 Ω at the antenna's center $(Z_0(0) = 100\ \Omega)$ to about 377 Ω at the antenna's aperture $(Z_0(X) = 377\ \Omega)$ over the UWB frequency range of 3.1–10.6 GHz. The characteristic impedances of the tapered sections are selected to produce minimum internal reflections along the antenna length (i.e., along the slot lines) over the entire UWB frequency range. Considering these conditions and assuming the maximum allowable input reflection coefficient $|R(0)|_{max} = 0.1$, the design parameter B can be derived as 1.53 from

(5.3). Considering the lower frequency limit of 3.1 GHz, from (5.3), the antenna half-length is then selected as $X = 0.6$ in.

The initial values of the characteristic impedances Z_0's of the tapered slot line at different locations along its length (x direction) are selected using (5.2) to produce minimum internal reflections for the antenna input signal over 3.1 to 10.6 GHz. The initial value of the terminating characteristic impedance at the aperture of the antenna is selected as the intrinsic impedance of free space (377 Ω). Since this impedance only works for spherical waves in free space, which is not the condition in our UWB antenna design, this impedance is probably not the optimum value for $Z_0(X)$ to achieve a smooth transition at the end of the antenna to the free space. Therefore, the EM simulator in Microwave Studio [15] was used to perform the time-domain EM simulations for the UWB uniplanar antenna and to optimize its structure to minimize reflections occurring at the open-end transition.

Another issue concerning the variations of the characteristic impedance Z_0 of the non-uniform slot lines is the implementation of the contour in the tapered slot lines. To provide a smooth transition that minimizes possible internal reflections along the antenna structure, the antenna should implement a gradual contour that avoids any abrupt transition in the shape across the entire antenna structure. For the non-uniform slotline transition sections close to the antenna center, the gap width is much smaller as compared to the associated metal widths, therefore the conventional characteristic-impedance formula of the slot line can be used to determine the gap width for the relatively smaller characteristic impedance. On the contrary, for the non-uniform slot line transition sections close to the open end of the antenna, the gap widths keep increasing, while the metal widths drop quickly, so the typical slot line calculation method cannot be applied to the transition structure anymore, instead the EM simulator is used to derive the accurate gap width for the specific metal widths and required characteristic impedance. Based on above-mentioned initial conditions, the variable characteristic impedances of the tapered slot line are determined from (5.2). Considering the abrupt variation of the slot edges around the open end of the antenna to accommodate the slot line's 377-Ω characteristic impedance, non-uniform steps are applied at points along the x-axis for the antenna structure, which are sparse around the antenna center and condensed round the open end, to maintain good variation accuracy. Through EM simulations with the EM simulator IE3D, the final optimized characteristic impedances $Z_0(x)$ of the tapered slot lines and their corresponding slot-line gap widths $G(x)$ and metal widths $W(x)$ (along the y-direction) of the UWB uniplanar antenna are derived, which are listed in Table 5.1 for each coordinate value along the x-axis.

To accommodate the investigation of the designed UWB uniplanar antenna's performance in both frequency and time domains, time-domain simulator Microwave Studio was used to simulate the performance of the designed UWB uniplanar antenna.

Figure 5.2 shows the simulated return loss of the UWB uniplanar antenna in frequency domain. The simulated results show more than 15-dB return loss over 3 to 12 GHz for the UWB uniplanar antenna. Figure 5.3 shows the reflection at the input of the antenna, where a Gaussian monocycle pulse with the 50% pulse width

Table 5.1 Characteristic impedances and corresponding dimensions of the UWB uniplanar antenna's tapered slot lines.
1 mil = 0.001 inch

x (mil)	$Z_0(x)$ (Ω)	$G(x)$ (mil)	$W(x)$ (mil)
10	100	30	750
58.8	101	32.4	747.4
117.1	103	37.2	740.6
174.2	107	52.8	725.7
229.6	113	78.6	702.9
282.8	123	108.2	675.5
333.3	139	145.6	641.6
380.6	151	190.2	601.8
424.3	165	242	556.1
463.8	178	313	498.8
498.9	193	392	435.7
529.2	208	477.8	367.5
554.3	226	565.6	297
574.2	240	654.8	224.7
588.7	251	748.4	149.3
597.1	257	839.8	74.5
600	260	930	0

Fig. 5.2 Simulated return loss of the designed UWB uniplanar antenna

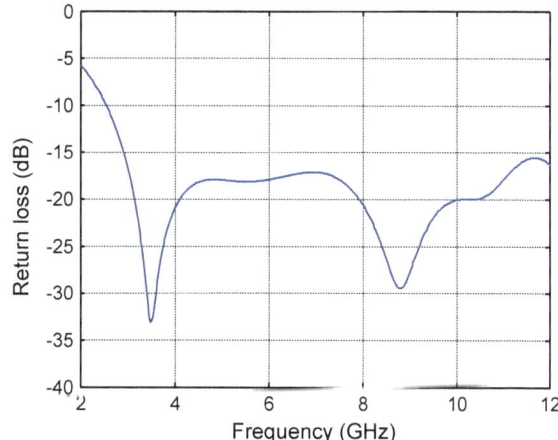

of 50 ps is used as the driving input signal. As can be seen, only small reflections occur, which confirm the good matching of the designed antenna for use in UWB impulse systems. Both the simulated frequency- and time-domain reflection validate the design of the UWB uniplanar antenna structure, particularly the optimized contour of the tapered slot lines, which plays the dominant role in the antenna's performance.

The transfer function of the UWB uniplanar antenna was also simulated with Microwave Studio at the position of 2 in. directly above the center of the antenna's surface. The simulated results are shown in Figs. 5.4 and 5.5. In Fig. 5.4, the

Fig. 5.3 Simulated input
reflection of the designed
UWB antenna in time-domain

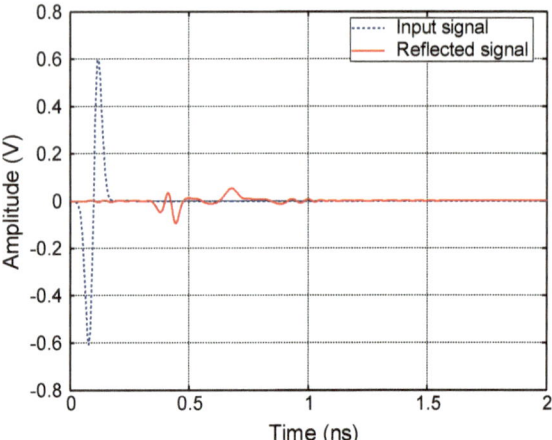

Fig. 5.4 Simulated
normalized amplitude of the
UWB uniplanar antenna's
transfer function

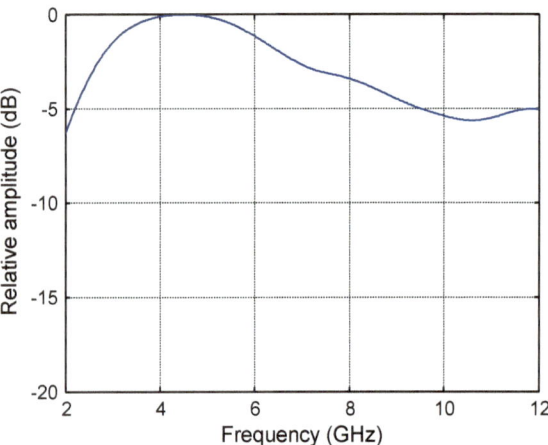

transfer function is normalized to facilitate understanding, and the simulated
normalized amplitude of the antenna's transfer function show the band-pass characteristic. The phase of the antenna's transfer function, as shown in Fig. 5.5,
indicates good linearity over the entire UWB band, implying minimum distortion to
the waveforms of the pulses transmitted or received by the antenna, which is crucial
for its use in UWB impulse systems.

   Figure 5.6 shows the simulated antenna gain of the UWB uniplanar antenna.
These results indicate that the maximum antenna boresight gain is normally 2.2 dBi
at 3.1 GHz.

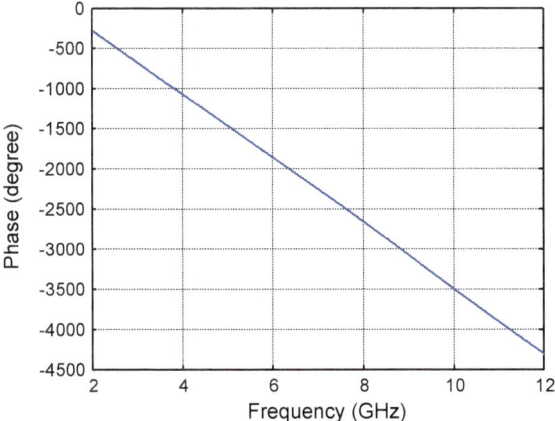

Fig. 5.5 Simulated phase of the UWB uniplanar antenna's transfer function

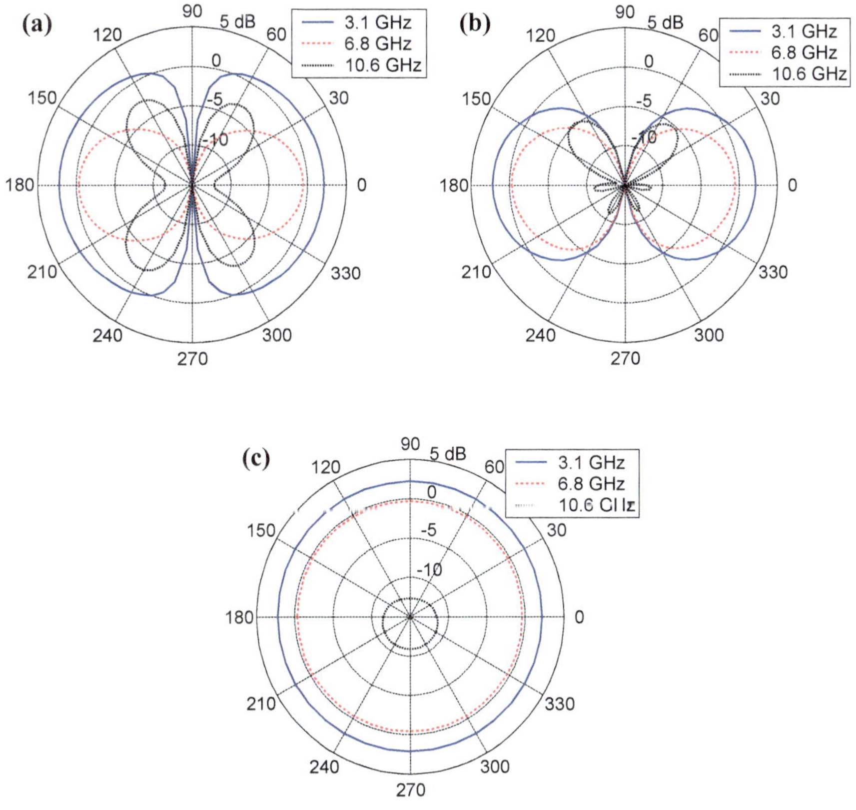

Fig. 5.6 Simulated patterns of the UWB uniplanar antenna: **a** *E*-plane (*x-y* plane), **b** *E*-plane (*y-z* plane), and **c** *H*-plane patterns at 3.1, 6.8, and 10.6 GHz. The coordinates (*x, y*) are shown in Fig. 5.1

5.3   UWB Uniplanar Antenna Fabrication and Test

Figure 5.7 shows a photograph of the UWB uniplanar antenna, including its aperture, SMA fixture, and feeding uniform CPW transmission line. The area occupied by the antenna aperture is 1.2 in × 1.5 in. It is a planar non-uniform-transmission-line-based traveling-wave antenna fabricated on a Duroid substrate having 0.025 in. thickness and relative dielectric constant of 10.5.

As mentioned earlier, for impulse UWB applications, as is considered here, the time-domain performance of the antenna is much more critical than its frequency counterpart. The antenna is used for transmitting or receiving UWB time-domain signals (impulse or monocycle pulses as addressed here), not multiple discrete frequency components in CW mode, making its time-domain measurement essential.

Figure 5.8 shows the measured and simulated results of the return loss in the frequency domain. The measured result shows more than 12-dB return loss over the entire 3.1–10.6 GHz UWB frequency band. As this return loss includes all effects from the designed antenna, CPW feed line, and SMA connector, it is impossible to derive the antenna's actual performance from the frequency-domain results.

On the contrary, it is relatively very easy to distinguish the antenna performance from other effects in the time domain, as will be shown subsequently. Furthermore, as the antenna is intended for radiating impulse or monocycle pulses, it is imperative to characterize it in time domain as discussed earlier. Figure 5.9 shows the measured and simulated time-domain reflectometry (TDR) response results in time domain for a 50-ps input impulse signal. It is apparent that, from 0 to 0.5 ns, the response corresponds to the effects of the SMA connector and CPW feed line. The response after 0.5 ns is caused by the antenna itself and, as can be seen, little reflection occurs with better than 18-dB return loss achieved for the antenna with

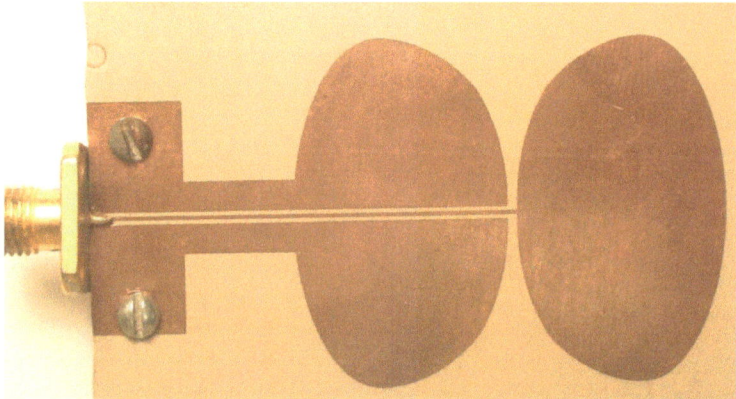

Fig. 5.7 Photograph of the UWB uniplanar antenna along with the 50-Ω CPW feed line and SMA connector (on the *left*)

Fig. 5.8 Measured and simulated return loss of the uniplanar UWB antenna

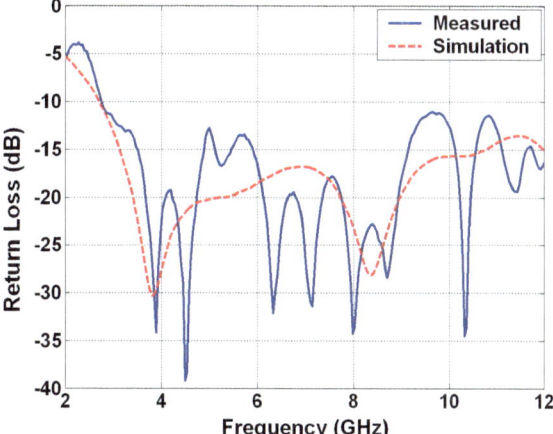

Fig. 5.9 Measured and calculated TDR responses of the uniplanar UWB antenna

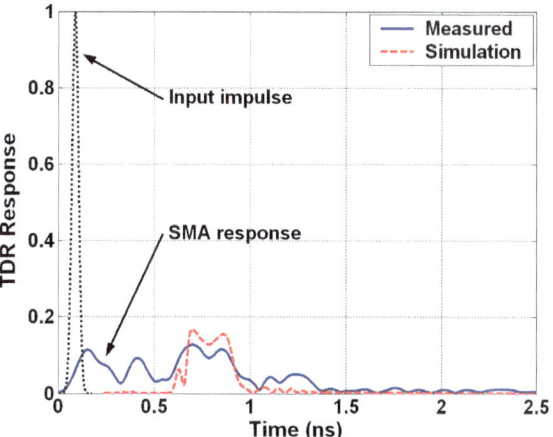

the driving impulse signal. The measured result matches very well with that simulated, which again confirms the antenna design. The TDR performance indeed demonstrates excellent time-domain behavior of the designed UWB uniplanar antenna, which is crucial for UWB time-domain impulse applications.

The results of of the designed UWB uniplanar antenna in both frequency and time domains demonstrate its usefulness for UWB applications. Moreover, the small size and uniplanar structure make the UWB uniplanar antenna very suitable for integration with the printed-circuit UWB transmitters and receivers designed and presented in the previous chapters.

5.4   Performance of UWB Transmit Prototype Integrating UWB Uniplanar Antenna and UWB Pulse Generator

In the final realization of the UWB system, both the CMOS UWB transmitter and receiver chips, designed and described previously, are integrated directly with the designed UWB uniplanar antenna. However, to show the feasibility and performance of an integrated UWB transmitter module, we integrated a packaged CMOS chip containing the previously designed tunable UWB monocycle-pulse generator [16] (besides other CMOS RFICs) with the UWB uniplanar antenna as a mockup and tested. This integration, as can be recognized, is not the optimum integration and is not implemented in the final UWB system. Figure 5.10 shows a photograph of the UWB transmit prototype integrating a packaged CMOS chip containing the CMOS tunable UWB monocycle-pulse generator [16] with the UWB uniplanar antenna. It is noted that the UWB monocycle-pulse generator used in the UWB transmit prototype is not the UWB monocycle-pulse generator described in Chap. 3, but another UWB monocycle-pulse generator that we previously reported in [16]. The CMOS chip is mounted directly onto the edge of the antenna without a feed line. The transmission line connecting the SMA connector and the CMOS chip, used for feeding the external 10-MHz clock signal, and the bias lines are etched onto the same board of the antenna.

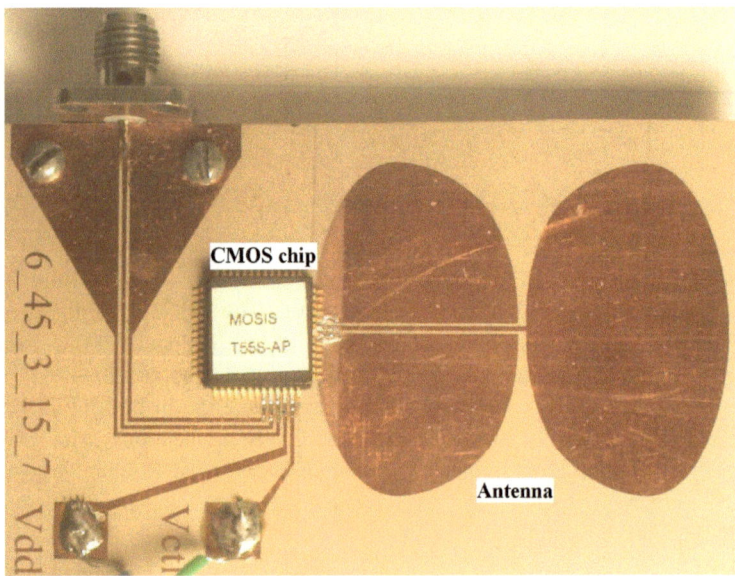

Fig. 5.10 Photograph of the UWB transmit prototype consisting of a packaged CMOS chip containing the CMOS tunable UWB monocycle-pulse generator integrated with the UWB uniplanar antenna

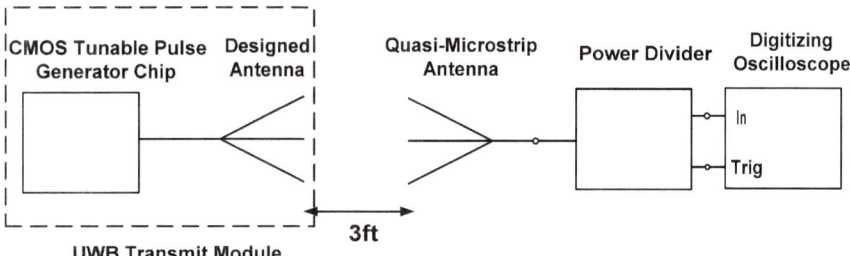

Fig. 5.11   Test setup for the pulse transmission measurement of the UWB transmit prototype

Figure 5.11 shows the block diagram of the test setup used for the pulse transmission measurement of the UWB transmit prototype. The microstrip Microstrip quasi-horn antenna operating from 0.2 to more than 20 GHz is used as the receiving antenna since it can produce faithfully the waveform of the received UWB signal. The UWB antenna of the UWB transmit prototype and the microstrip quasi-horn antennas face each other and spaced 3-ft apart. The pulse received by the microstrip quasi-horn antennas is fed into a power divider and displayed in a 50-GHz digitizing oscilloscope.

Figure 5.12 shows the pulse signals received from the tunable impulse signals, shown in Fig. 5 of [16], transmitted by the UWB transmit module. The pulse-duration tunability is clearly visible in the received pulses. As can be seen, the received signals are monocycle pulses with pulse duration tunable from 160 to 350 ps. The resultant monocycle waveform is due to the differential function of the designed antenna. The received pulses maintain good symmetry with no serious distortion and ringing.

Figure 5.13 shows the received pulse signal corresponding to the monocycle pulse signals, shown in Fig. 6 in [16], transmitted by the UWB transmit module. The received pulse also has tunable durations. All the received signals have shape

Fig. 5.12   Measured received signals of the impulses transmitted by the UWB transmit prototype for different control voltages

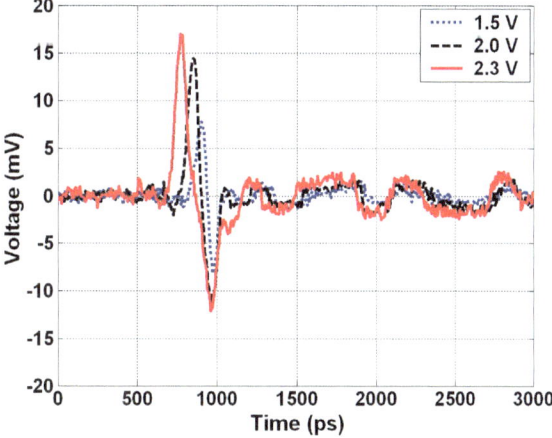

Fig. 5.13 Measured received signals of the monocycle pulses transmitted by the UWB transmit prototype for different control voltages

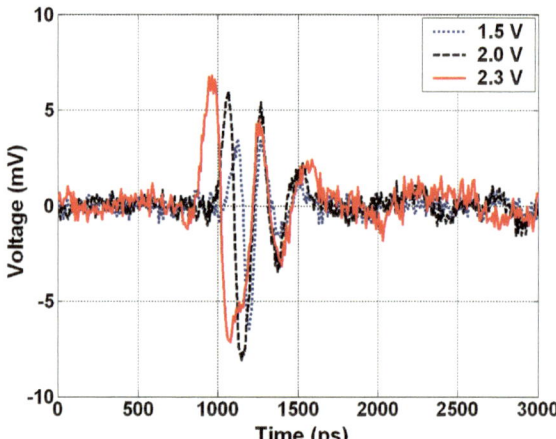

similar to the first derivative of the monocycle pulses, as expected from the designed antenna. Both the measured impulse and monocycle-pulse transmission results clearly demonstrate the workability of the developed CMOS-based tunable UWB transmit prototype.

5.5   Summary

This chapter covers the design and performance of a low-cost, compact UWB uniplanar antenna operating across the entire UWB of 3.1–10.6 GHz, which can be easily integrated with the designed CMOS UWB transmitter and receiver. Especially, various impulse time-domain measurements are conducted to demonstrate the performance and suitability of the designed UWB uniplanar antenna for transmitting and receiving impulse signals that contain all frequency components from 3.1 to 10.6 GHz, which are essential for UWB impulse systems. The UWB uniplanar antenna is further integrated with the previously developed 0 tunable pulse generator to form the UWB transmit prototype. This UWB transmit prototype can transmit monocycle pulses with pulse duration tunable from 160 to 350 ps with the impulses supplied from the integrated pulse generator. The received pulses maintain good symmetry with no serious distortion and ringing. The UWB transmit prototype can also transmit signals having shape similar to the first derivative of the monocycle pulses provided by the integrated pulse generator. Both the impulse and monocycle-pulse transmission results clearly demonstrate the workability of the CMOS-based tunable UWB transmit prototype for UWB impulse systems.

References

1. H.G. Schantz, Introduction to ultra-wideband antennas. In *Conference on Ultra Wideband Systems and Technologies,* pp. 1–9 November 2003
2. A.G. Yarovoy, B. Sai, G. Hermans, P.V. Genderen, L.P. Lighart, A.D. Schukin, and I.V. Kaploun, Ground penetrating impulse radar for detection of small and shallow-buried objects. In *IGARSS'99 Proceedings of IEEE 1999 International Geoscience and Remote Sensing Symposium, 1999,* vol. 5, pp. 2468–2470
3. D.J. Daniels, Short pulse radar for stratified lossy dielectric layer measurement. In *IEEE Proc* vol. 127, (pt. F, no. 5), pp. 384–388, Oct.1980
4. J.A.N. Noronha, T. Bielawa, C.R. Anderson, D.G. Sweeney, S. Licul, and W.A. Davis, Designing antennas for UWB systems. Microwave & RF (2003)
5. J.D. Cermignani, R.G. Madonna, P.J. Scheno, J. Anderson, Measurement of the performance of a cavity backed exponentially flared TEM horn. Proc. SPIE Utrawideband Radar **1631**, 146–154 (1992)
6. J. Park, C. Nguyen, An ultra-wideband microwave radar sensor for characterizing pavement subsurface. IEEE MTT-S Int. Microwave Sym. Digest **2**, 1143–1146 (2003)
7. C. Nguyen, J.S. Lee, and J.S. Park, Novel Ultra-Wideband Microstrip Quasi-Horn Antenna. Electr. Lett. **37**(12), pp. 731–732 (2001)
8. J.S. Lee, C. Nguyen, and T. Scullion, A novel compact, low-cost impulse ground penetrating radar for nondestructive evaluation of pavements. In *IEEE Transcation on Instrumentation and Measurement,* Vol. IM-53, pp. 1502–1509, Dec. 2004
9. J.W. Han, C. Nguyen, Development of a tunable multi-band UWB radar sensor and its applications to subsurface sensing. IEEE Sens. J. **7**(1), 51–58 (2007)
10. J. Han and C. Nguyen, Investigation of Time-Domain Response of Microstrip Quasi Horn Antennas for UWB Applications. In IEEE Electr. Lett **43**(1) pp. 9–10 Jan. 2007
11. H.G. Schantz, Bottom fed planar elliptical UWB antennas. In IEEE *Confererence on Ultra Wideband Systems and Technologies,* (Nov. 2003), pp. 219–223
12. E.A. Theodorou, M. R. Gorman, P. R. Rigg, and F. N. Kong, Broadband pulse-optimised antenna. *IEE Proc.,* vol. 128, (pt. H, no.3), pp. 124–130, June 1981
13. R.P. Hecken, A near-optimum matching section without discontinuities. IEEE Trans. Microwave Theory Tech. **20**(11), 734–739 (1972)
14. Open Cavity Plastic Packages, *MOSIS foundry* (Marina Del Rey, CA, 2005)
15. CST Microwave Studio, *CST of America Inc* (Wellesley Hills, MA, 2005)
16. M. Miao and C. Nguyen, On the Development of an Integrated CMOS-Based UWB Tunable–Pulse Transmit Module, In *IEEE Trans. on Microwave Theory and Technique,* Vol. MTT-54, No. 10, Oct. 2006, pp. 3681–3687

Chapter 6
Summary and Conclusions

In the last five chapters, this book has covered the theory, analysis and design of the UWB impulse transmitters and receivers, and their constituent components, along with the design of a UWB uniplanar antenna for UWB impulse systems operating from 3.1 to 10.6 GHz for short-range wireless communications and sensing. The same UWB transmitters and receivers can also be used for longer ranges provided that the pulse generators are designed with higher pulse amplitudes. Specifically, the following materials are addressed in this book:

Chapter 1 introduces UWB impulse systems for wireless communications and sensing.

Chapter 2 covers the fundamentals of UWB impulse systems operating across or within the unlicensed UWB frequency band of 3.1–10.6 GHz. It provides the essence of UWB impulse systems including the spectrum mask, advantages and applications of UWB impulse systems, UWB impulse signals including Gaussian impulse, doublet pulse and monocycle pulse, modulations including PPM, PAM, OOK and BPSK, UWB impulse transmitters and receivers, and UWB antennas.

Chapter 3 presents the design of CMOS UWB tunable sub-nanosecond impulse and monocycle-pulse BPSK transmitters. The CMOS UWB impulse transmitter produces positive and negative tunable impulse signals corresponding to the high- and low-level modulation signals from the BPSK modulator operated in accordance with the "1" or "0" digital data information, respectively. The positive and negative impulse signals have amplitudes of 0.8 V and 0.6 V with tunable pulse widths from 100 to 300 ps, respectively. The CMOS UWB monocycle-pulse transmitter produces monocycle pulse signals of opposite polarities having peak-to-peak amplitudes of about 0.6–0.8 V and tunable pulse widths between 100 and 300 ps. This chapter also covers the design of the CMOS impulse generator and BPSK modulator used in the CMOS UWB impulse and monocycle-pulse transmitters. The CMOS impulse generator can generate 0.95–1.05 V peak-to-peak Gaussian-type impulse signal with 100–300 ps tunable pulse duration. Moreover, this chapter includes the design of a CMOS monocycle pulse generator, which can produce 0.7–0.75 V peak-to-peak monocycle pulse with 140–350 ps tunable pulse duration.

C. Nguyen and M. Miao, *Design of CMOS RFIC Ultra-Wideband Impulse Transmitters and Receivers*, SpringerBriefs in Electrical and Computer Engineering, DOI 10.1007/978-3-319-53107-6_6

Chapter 4 presents the design of a CMOS UWB impulse receiver and its core components of UWB LNA and correlator. The UWB LNA employs the cascoded common-source inductively degenerated topology with an extended UWB ladder matching network. The shunt-peaking topology is also implemented for the LNA to further improve its performance at the high-frequency end. The LNA utilizes the PGS spiral inductors designed for high Q to further improve the LNA's performance. The return losses of the LNA with the source-follower buffer for both the input port and output port are better than 10 dB over the entire UWB band of 3.1–10.6 GHz. The reverse isolation of the LNA of more than 40 dB is also achieved over the UWB range. The UWB LNA achieves the maximum gain of 12.4 dB over the UWB band. The UWB LNA possesses an average NF of 5.8 dB over the entire UWB band. The UWB multiplier of the UWB correlator is based on the transconductor multiplier structure with the central component of CMOS programmable transconductors. It converts the input voltage signal into differential current to realize the multiplication. The shunt-peaking topology is also applied at the output, which achieves the pole-zero cancellation and extends the multiplier bandwidth from 2 to 10 GHz for un-load situation, and 7 GHz for buffer-load condition. Finally, the UWB LNA is integrated with the UWB multiplier and template pulse generator to form the UWB receiver, which demonstrates its ability to receive and convert sub-nano-second pulse signals to baseband signals for UWB impulse systems.

Chapter 5 covers the design and performance of a low-cost, compact UWB uniplanar antenna operating across the entire UWB of 3.1–10.6 GHz, which can be easily integrated with the designed CMOS UWB transmitter and receiver. Especially, various impulse time-domain measurements are conducted to demonstrate the performance and suitability of the designed UWB uniplanar antenna for transmitting and receiving impulse signals that contain all frequency components from 3.1 to 10.6 GHz, which are essential for UWB impulse systems. The UWB uniplanar antenna is further integrated with the previously developed CMOS tunable pulse generator to form the UWB transmit prototype. This UWB transmit prototype can transmit monocycle pulses with pulse duration tunable from 160 to 350 ps with the impulses supplied from the integrated pulse generator. The received pulses maintain good symmetry with no serious distortion and ringing. The UWB transmit prototype can also transmit signals having shape similar to the first derivative of the monocycle pulses provided by the integrated pulse generator. Both the impulse and monocycle-pulse transmission results clearly demonstrate the workability of the CMOS-based tunable UWB transmit prototype for UWB impulse systems.

Successful development of the UWB impulse transmitter and receiver in CMOS technology and UWB uniplanar antenna, along with the demonstration of the UWB transmit prototype integrating the UWB pulse generator and UWB uniplanar antenna, demonstrate their workability and usefulness for UWB impulse systems for various UWB applications including UWB communication systems, sensors, and radars. This successful development also paves the way for more advanced UWB transceiver modules integrating CMOS transceiver chips directly with antennas.

Although the UWB transmitter, receiver and antennas developed in this book are used specifically for the UWB impulse system, they can also be used for other UWB time-main applications as well as in other UWB systems. Particularly, their analyses and design techniques would enable engineers to design UWB components and systems for a variety of specifications and applications.

We wish to emphasize that, even though this book is relatively concise, it contains sufficient practical and valuable information that should enable the readers to successfully design and measure UWB impulse transmitters, receivers, and antennas relatively easy for their own use in many applications.

Lastly, we wish to make several final remarks that could be useful for designers of UWB systems wanting to apply some of the designs described in this book to further improve the performance of UWB components and subsystems, and hence UWB systems.

The performance of the UWB impulse transmitter can be improved by rejecting the noise coupled through the silicon substrate, which is substantial due to the conductive nature of the silicon substrate, by implementing the common-mode rejection. To that end, differential circuits would be a better choice for the UWB impulse transmitter design. Furthermore, both NOR and NAND gate blocks should be selected to replace the CMOS inverter at the BPSK modulator input to achieve symmetric positive and negative impulse and monocycle pulses in the BPSK-integrated tunable pulse generator. In addition, better delay-cell type should be selected to achieve a broader linear tuning range versus tuning voltage variation.

Similarly, the single-ended LNA should be replaced with a differential structure in the UWB impulse receiver along with a differential-type UWB multiplier. This will make the UWB impulse receiver capable of common-mode rejection, hence suppressing the noise coupled through the silicon substrate, which cannot otherwise be avoided with single-ended circuits.

Bibliogarphy

[1] N.R. Mahapatra, A. Tareen, S V. Garimella, in Comparison and analysis of delay elements. *The 2002 45th Midwest Symposium on Circuits and Systems, MWSCAS-2002,* vol. 2, August 2002. pp. 473–476

[2] A.G. Yarovoy, B. Sai, G. Hermans, P.V. Genderen, L.P. Lighart, A.D. Schukin, and I.V. Kaploun, Ground penetrating impulse radar for detection of small and shallow-buried objects, in *IGARSS '99 Proceedings of IEEE 1999 International Geoscience and Remote Sensing Symposium,* 1999, vol. 5, pp. 2468–2470.

[3] D.J. Daniels, Short pulse radar for stratified lossy dielectric layer measurement. in *IEE Proc*eedings, vol. 127, Oct. 1980. pt. F, no. 5, pp. 384–388

[4] J.D. Cermignani, R.G. Madonna, P. J. Scheno, and J. Anderson, Measurement of the performance of a cavity backed exponentially flared TEM horn, in *Proceedings of SPIE: Utrawideband Radar,* vol. 1631, May 1992. pp. 146–154

[5] J. Park and C. Nguyen, An ultra-wideband microwave radar sensor for characterizing pavement subsurface, in *IEEE MTT-S International Microwave Symposium Digest,* vol. 2, Jun. 2003. pp. 1143–1146

[6] C. Nguyen, J.S. Lee, and J.S. Park. Novel Ultra-Wideband Microstrip Quasi-Horn Antenna. Electron. Lett. Vol. **37**(12), 731–732 (2001)

[7] J. Han and C. Nguyen. Investigation of Time-Domain Response of Microstrip Quasi Horn Antennas for UWB Applications. IEE Electron. Lett. Vol. **43**(1) 9–10 (2007)

[8] J. R. *Andrews UWB signal sources & antennas.* Application Note AN-14, (Picosecond Pulse Labs, Boulder, 2003)

[9] H.G. Schantz. in Bottom fed planar elliptical UWB antennas IEEE Conference on Ultra Wideband Systems and Technologies, Nov. 2003, pp. 219–223.

[10] E.A. Theodorou, M.R. Gorman, P.R. Rigg, and F.N. Kong, in Broadband Pulse-Optimised Antenna, *IEE Proc*eedings, vol. 128, Jun. 1981, pt. H, no.3, pp. 124–130

[11] R P. Hecken. A near-optimum matching section without discontinuities. *IEEE Trans. Microwave Theory and Tech.,* vol. **20**(11) 734-739 (1972)

© The Author(s) 2017
C. Nguyen and M. Miao, *Design of CMOS RFIC Ultra-Wideband Impulse Transmitters and Receivers*, SpringerBriefs in Electrical and Computer Engineering, DOI 10.1007/978-3-319-53107-6

[12] *Open Cavity Plastic Packages*. MOSIS foundry, (Marina Del Rey, Los Angeles, 2005)

[13] *CST Microwave Studio*. CST of America Inc., (Wellesley Hills, Boston)

Index

© The Author(s) 2017
C. Nguyen and M. Miao, *Design of CMOS RFIC Ultra-Wideband Impulse Transmitters and Receivers*, SpringerBriefs in Electrical and Computer Engineering, DOI 10.1007/978-3-319-53107-6